Forest and Plant Bioproducts Division 2021

Held at the 2021 AIChE Annual Meeting

Boston, Massachusetts, USA and Online
7 - 11 November and 15 - 19 November 2021

ISBN: 978-1-7138-5711-2

Printed from e-media with permission by:

Curran Associates, Inc.
57 Morehouse Lane
Red Hook, NY 12571

Printed with permission by Curran Associates, Inc. (2022)

For permission requests, please contact AIChE
at the address below.

AIChE
120 Wall Street, FL 23
New York, NY 10005-4020

Phone: (800) 242-4363
Fax: (203) 775-5177

www.aiche.org

Additional copies of this publication are available from:

Curran Associates, Inc.
57 Morehouse Lane
Red Hook, NY 12571 USA
Phone: 845-758-0400
Fax: 845-758-2633
Email: curran@proceedings.com
Web: www.proceedings.com

TABLE OF CONTENTS

*** Formatting Issue - Best Available Paper/Graphic ***

622660 Towards a New Understanding of the Retro-Aldol Reaction for Oxidative Conversion of Lignin to Aromatic Aldehydes and Acids

Abstract:

Lignin is the most abundant natural biopolymer after cellulose and accounts for 30% of non-fossil carbon on earth. The retro-aldol reaction is one of the key steps involved in the oxidative conversion of lignin to aromatic aldehydes and acids. In principle, the retro-aldol reaction can proceed in the absence of oxygen. In this work, a new approach based on the influence of oxygen on the oxidation of lignin was investigated. The effect of reaction chemistry, time, temperature, and lignin feedstock play a key role on the yield of aromatic aldehydes and acids. At 140°C, oxidation of softwood Lignoboost kraft lignin for 40 minutes results in combined maximum yield of 5.17% w/w of vanillin and vanillic acid. In comparison, using the new approach in which oxygen was charged for only 20 minutes during the 40 minute reaction improved this yield considerably to 6.95%. Further, yield improvement was obtained when applying this approach to different lignin feedstocks. Oxidation also increased the carboxyl content of lignin from 0.49 mmol/g to 1.41 mmol/g which represents consequential improvement. The current study provides further evidence showing that the oxidation reaction is a crucial pathway for lignin valorization.

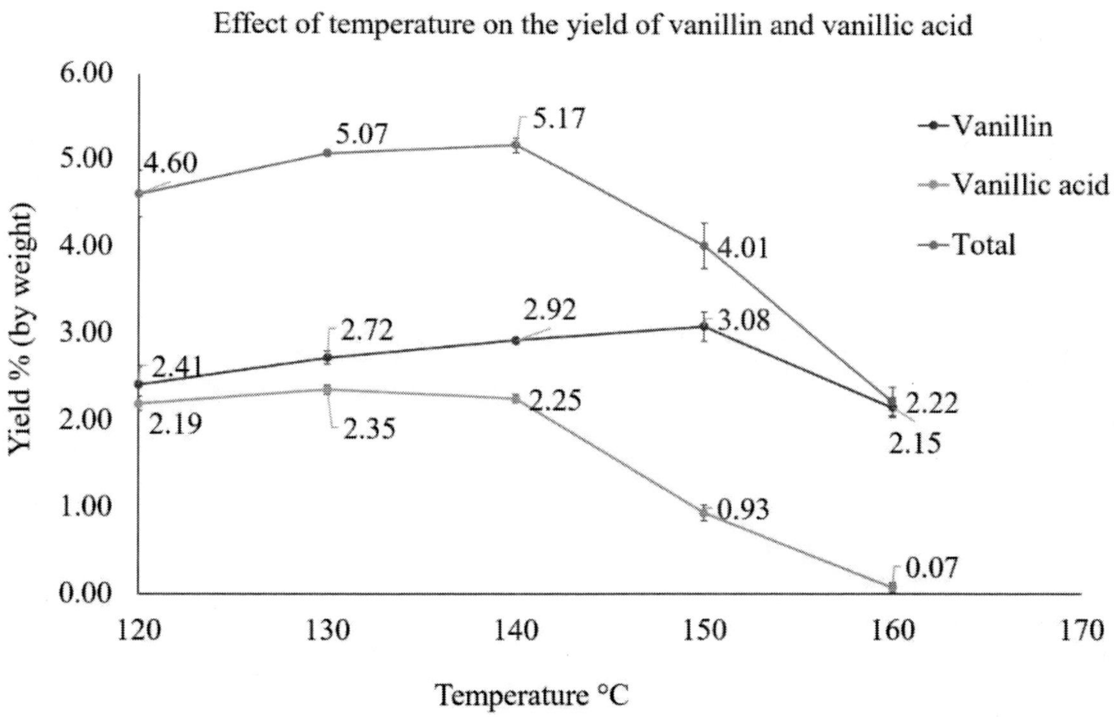

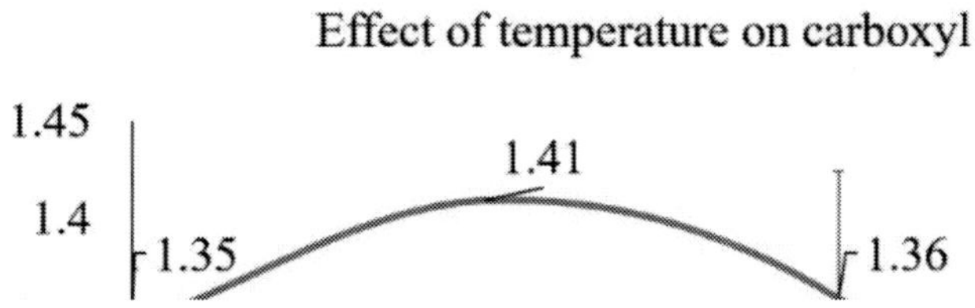

*** Formatting Issue - Best Available Paper/Graphic ***

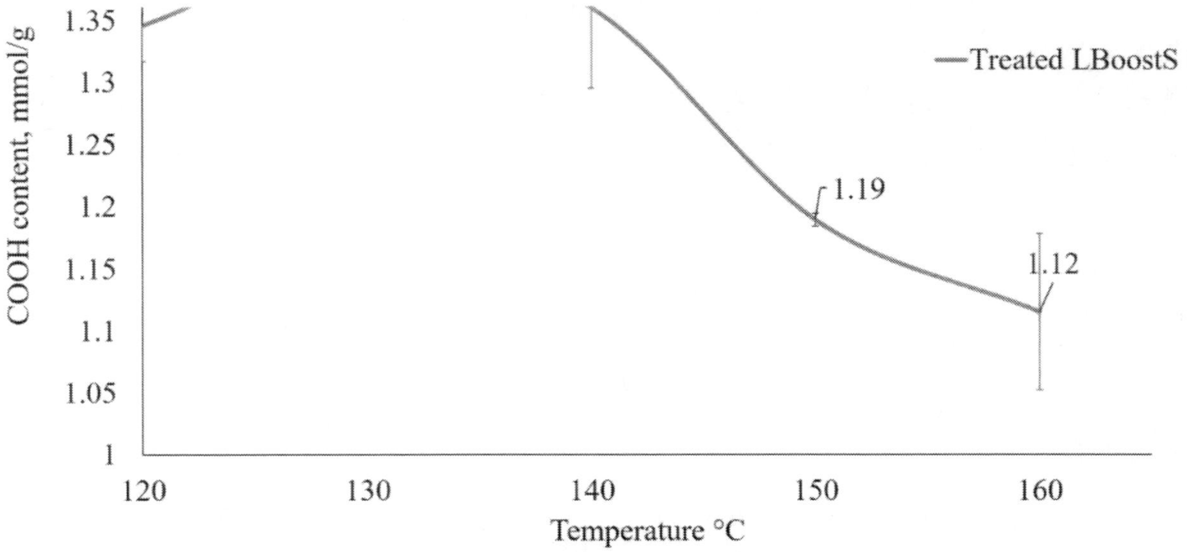

Effect of time on the yield of vanillin and vanillic acid

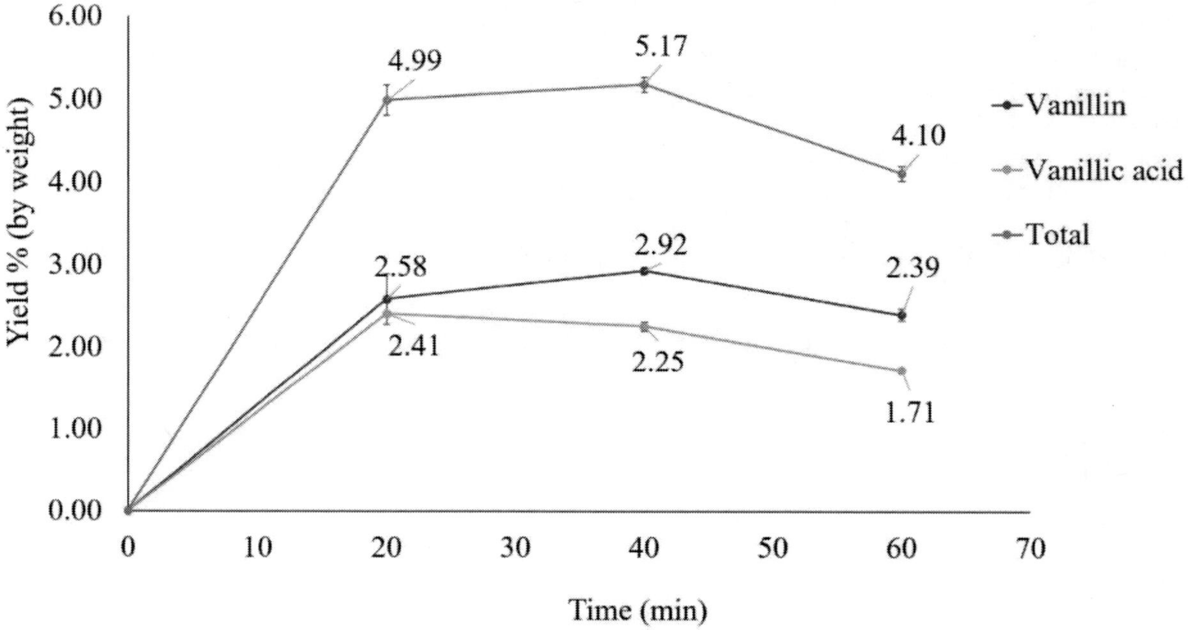

Effect of controlling the amount of oxygen charged during the reaction on the yield of vanillin and vanillic acid

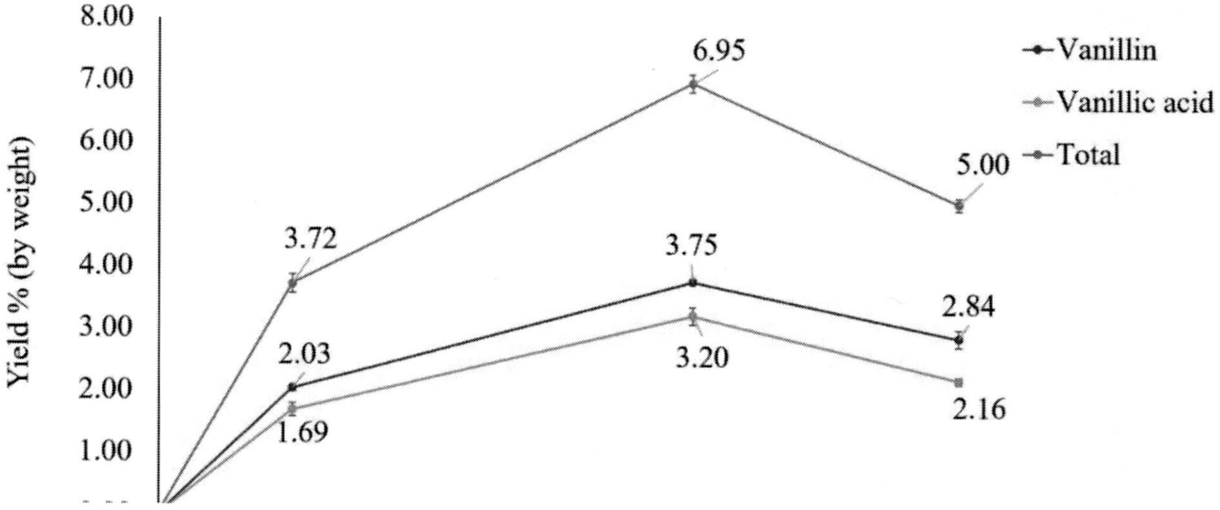

*** **Formatting Issue - Best Available Paper/Graphic** ***

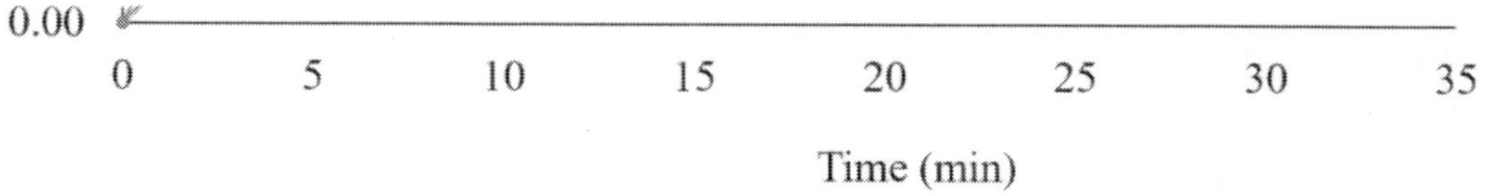

Effect of controlling the amount of oxygen charged during the reaction on the yield of vanillin and vanillic acid from different lignin feedstocks

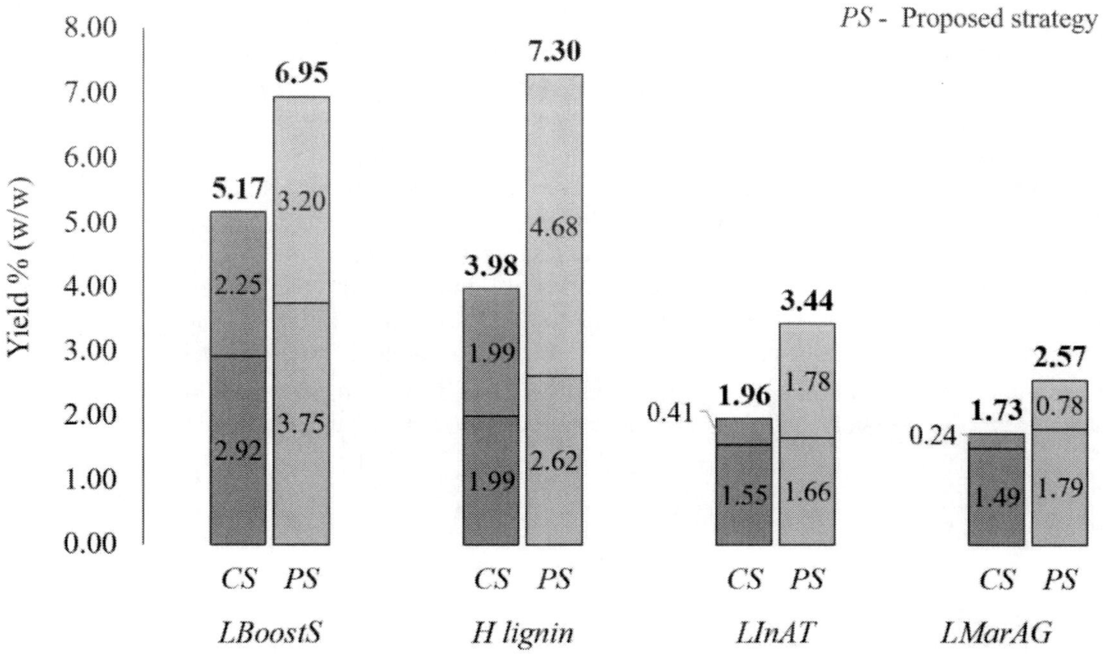

628460 Lignin Depolymerization in a Molten Salt Hydrate (lithium bromide trihydrate)

Abstract:

A molten salt hydrates (MSH) is a concentrated aqueous solution of an inorganic salt, in which the molar ratio of water to salt is close to the coordination number of the salt cation. Because of their unique properties, in particular the ability to dissolve cellulose, some MSHs are promising solvents and reaction media for biomass processing. For example, LiBr molten salt hydrate (~61 wt% LiBr aqueous solution) has been successfully used in hydrolysis of cellulose and lignocellulosic biomass, lignin quantitation, isomerization of glucose, furan-based chemicals from biomass, synthesis of oligosaccharides, and fabrication of nanocrystals, films, and aerogels from cellulose and biomass in our previous studies. It was also observed that lignin was extensively depolymerized when biomass was treated in the LiBr system. This presentation is to report our recent investigations into the depolymerization mechanisms of lignin in the LiBr system. The studies revealed that all the ether bonds of lignin in methoxyl, β-*O*-4, β-5, and β-β structures except for the 4-*O*-5 bond in diphenyl structure could be cleaved in the LiBr system. The ether bonds were cleaved via the S_N2 substitution pathway except for β-*O*-4 bond. The β-*O*-4 bond was primarily cleaved via the benzyl cation and enol ether intermediates, leading to Hibbert's ketones. Besides, some β-*O*-4 structures were transformed into benzodioxane structures, which were stable in the LiBr system.

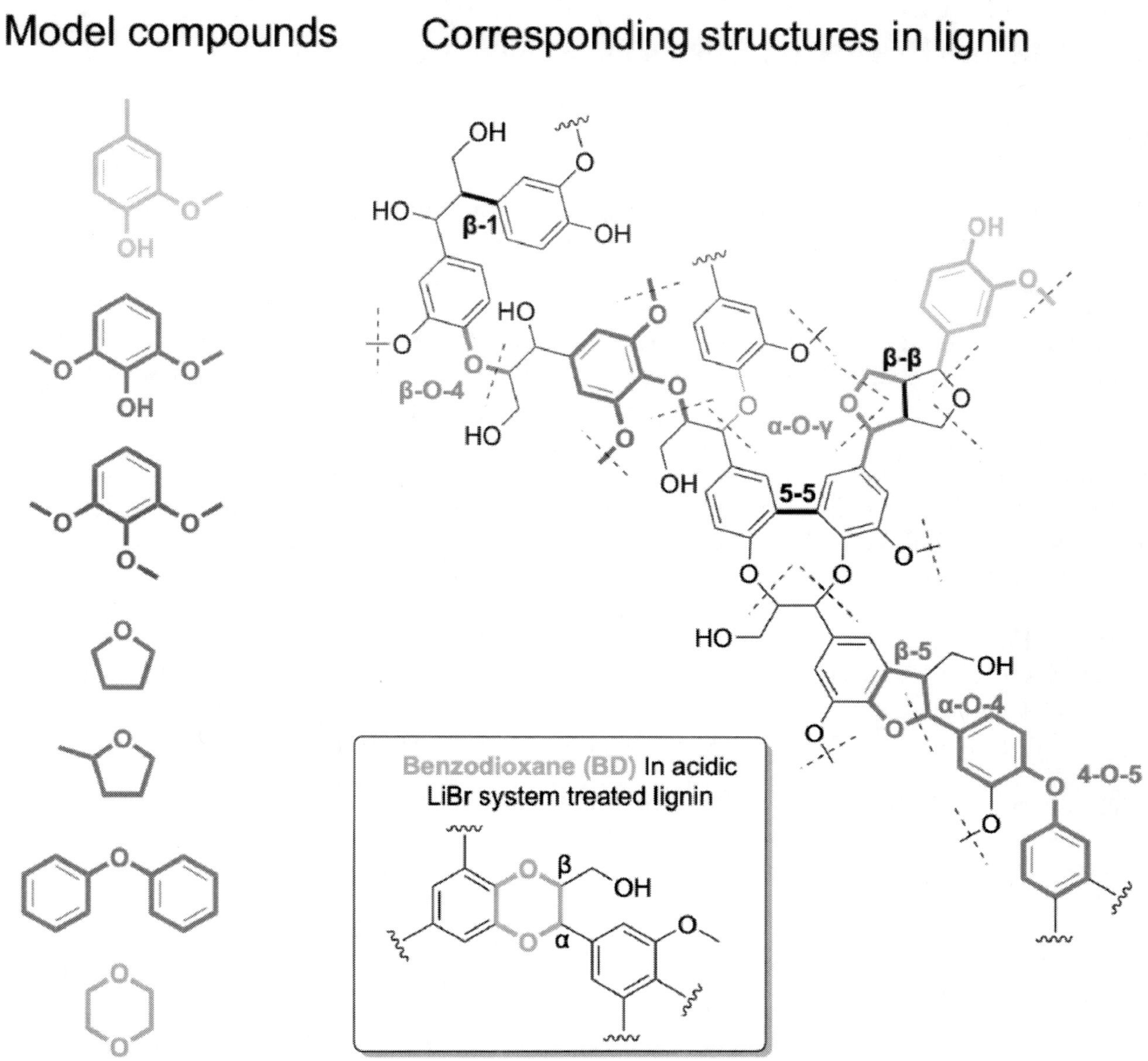
Model compounds
Corresponding structures in lignin
β-1
β-O-4
α-O-γ
β-β
5-5
β-5
α-O-4
4-O-5
Benzodioxane (BD) In acidic LiBr system treated lignin

631322 Direct Catalytic Conversion of Lignocellulosic Biomass into Liquid Paraffins and Aromatics

Abstract:

Today, gasoline is the major economic driver for petroleum refineries and its production accounts for nearly 20% of the total volume of products derived from a barrel of crude oil. However, the market share occupied by gasoline consumption is expected to decline 1% annually through 2050 and the petrochemical industry will become more dependent on the production of non-combustible petrochemicals, jet fuel and marine diesel due to increased car fuel efficiency, emerging electric and hydrogen fuel cell vehicle utilization. Thus, according to the BP Energy Outlook 2020, biofuels are expected to play a major role for the aviation and marine industries, but they also can play a role in accelerating the displacement of petroleum-based gasoline and non-combustible fuels, which could be substituted by the exact bio-based equivalents if there are biomass conversion technologies that enable such potential. Herein, a zeolite-supported metal catalyst in a presence of a hydrocarbon solvent was used to directly convert lignocellulosic biomass into a range of gaseous and liquid hydrocarbons at 280 Celsius via hydrodeoxygenation. Biomass conversion and carbon yields higher than 90% were obtained with a solvent to biomass ratio of 3:1 and 8h of residence time. C5-C7 alkanes (61% of total carbon) and C3-C4 alkanes (14% of total carbon) along with lower level of aromatics (10% of total carbon) were the main products generated. The produced alkanes can be fractionated and used directly in many applications today or subjected to further oxidative coupling followed by oligomerization to produce carbon range jet fuel and diesel.

626996 *Influence of Dilute Acid Pretreatment and Lignin Extraction Conditions on Lignin and Phenol Formaldehyde Resin Properties*

Abstract:

The type of feedstock used, pretreatment employed, and method of lignin extraction/purification all influence the properties of lignins extracted from a biorefinery process. Structural and chemical properties of a specific lignin such as phenolic hydroxyl content and molecular weight affect the ability of a lignin to be incorporated into phenol formaldehyde (PF) resins for wood adhesives, and the subsequent material properties of the formulated adhesives. Our previous studies have shown that lignin generated from NaOH extraction of dilute acid pretreated corn stover can be used to formulate PF resins with 100% of phenol replaced which exhibit material properties that exceed industrial resin standards. We have previously hypothesized that the improved lignin incorporation in this study is largely due to high amounts of *p*-coumarate (*p*CA) present within grass lignins presenting additional phenolic hydroxyls for reaction with formaldehyde. In a previous study we have identified dilute acid pretreatment conditions that result in high hydrolysis yields while retaining high *p*CA levels within the biomass. In this study we expand on those finding by using corn stover to investigate a wide range of dilute acid pretreatment conditions, varying enzymatic hydrolysis conditions, and two different lignin extraction processes using sodium hydroxide and formic acid. These varied conditions were used to create a panel of lignins with varying properties which were completely characterized and the most promising lignins were chosen for scaled up production and PF resin adhesive formulation. Using optimized dilute acid pretreatment conditions and formic acid extraction we were able to formulate lignin-formaldehye resins which exceeded properties of previously formulated resins.

629733 Fabrication of Hydrogel/Aerogel from Recycled Biomass

Abstract:

Hydrogel is a soft matter of a three-dimensional polymer network swollen by a large amount of water. Dependent on polymers, fabrication method, and functionalization, a hydrogel can have different properties such as stretchability, compressibility, water retention, pH sensitivity, and conductivity. Hydrogels have been widely utilized in the fields such as biomaterials, agriculture, and water purification. Aerogel is an ultralight material derived from a gel (in many cases, a hydrogel), where the liquid component in the gel has been replaced with a gas. In general, aerogels have low density, high porosity, and many unique properties in the meantime. Current hydrogels and aerogels are mainly fabricated from synthesized polymers or cross-linked by additional chemical agents. With the increasing demand for green material and sustainable processes, more attention has been drawn to developing hydrogels and aerogels from natural and renewable materials.

Cellulose is an abundant natural polymer on earth and has been used for fabricating hydrogels and aerogels. One of the cellulose-based hydrogel/aerogel fabrication approaches is the dissolution-regeneration process. Diverse solvent systems such as LiCl/dimethylacetamide (DMAc), *N*-methylmorpholine-*N*-oxide (NMMO), ionic liquids (ILs), and alkali/urea (or thiourea) aqueous systems have been used to fabricate cellulose-based hydrogels/aerogels in previous studies. Compared with the above solvents, inorganic salt solvents have lower vapor toxicity and relatively lower price. In addition, they are easy to use and recycle.

Herein, we report a facile route to fabricate cellulose hydrogel and aerogel directly from recycled biomass using zinc chloride salt hydrate as a cellulose solvent. Zinc chloride is a well-known cellulose swelling and dissolving agent. The cellulose regeneration process is simple, rapid, and under mild conditions to form a hydrogel. The aerogel obtained after the lyophilization of the hydrogel showed high stiffness and lightweight. The fabricated hydrogel and aerogel were characterized using dynamic mechanical analysis (DMA), scanning electron microscope (SEM), differential scanning calorimetry (DSC), and Fourier-transform infrared spectroscopy (FTIR). This study could pave a new way to valorize the recycled biomass.

630837 Assessment of Tribological Properties and Oxidative Stability of Bio-Based Lubricant from Pequi Oil (CARYOCAR BRASILIENSIS)

Abstract:

Bio-based lubricants are environmentally preferred to petroleum-based lubricants because they have greater biodegradability and are less toxic. Other advantages include very low volatility due to the high molecular weight and excellent viscosity properties with temperature variations. In this study, samples using pequi oil (*Caryocar Brasiliensis*) were obtained by hydrolysis and esterification reactions, using 2-ehtylhexanol. The oxidative stability was measured using Rancimat technique at 110°C. The induction period of 5.44±0.19 h was observed for the synthesized samples. Tribological tests in a four-ball configuration were performed to evaluate the lubricating behavior. A comparison of the Stribeck curve of the bio-based sample with that of a mineral hydrotreated basestock oil showed that the samples from pequi oil had a lower friction coefficient at all the sliding speeds studied. The friction coefficient of the bio-based sample was 30.7% lower than mineral hydrotreated basestock oil. The wear scar diameters were 0.371 mm for the bio-based samples and 0.195 mm for mineral oil. These results indicate that a bio-based samples obtained from pequi oil may be potentially used as a basestock oil since it presents a satisfactory tribological performance.

629760 Effect of Recycling HTC Process Liquid on Hydrochar Morphology and Its Corresponding Performance on Dye Adsorption

Abstract:

The dying process industries release many chemicals through the polluted wastewater that result in the death of aquatic life, ruining of soil and poisoning of drinking water. Henceforth, this study addresses the adsorption of dye from the wastewater through various quality hydrochars and optimize the adsorption process. Hydrochars are produced via Hydrothermal Carbonization (HTC) which is a thermochemical process that converts wet waste biomass into effective adsorbent. HTC has been successfully reported for wastes including but not limited to food waste, sewage sludge, dairy manure, municipal solid wastes, and agricultural wastes. Despite producing huge quantity of liquid in the reaction products, the hydrochar has received attention to date for its ability to act as an adsorbent. However, the HTC process liquid remains a concern for potential commercialization due to its high disposal costs and environmental pollution potentials. Therefore, this study focused on the recyclability of the process liquid to the HTC reaction to produce different quality hydrochars, to enhance the sorption capacity and further minimize the waste disposal cost. The consequence of preceding process liquid recycle at each HTC of corn stover for ten cycles with a focus on the char characteristic was studied. The HTC was done for 230 °C with a fixed residence time of 30 mins. The resulting chars were characterized through mass yield, ultimate analysis, and surface morphology (BET surface area, pore size, pore volume), pH, pH at point of zero charge (pH_{PZC}) and surface functionalities to evaluate their dye sorption capacity and two adsorption isotherm models (Langmuir and Freundlich) were fitted at 4, 20 and, 36 °C. Finally, a comparative analysis was done between the resulting chars and the optimum recycle number was found out in terms of adsorption performance.

627461 Heterologous Gene Expression Yields Higher Polyhydroxybutyrate Production in *Paraburkholderia Sacchari*

Abstract:

Recent research endeavors have turned to generating useful chemicals from biological platforms as an environmentally responsible alternative to non-sustainable sources. One such chemical is polyhydroxybutyrate (PHB), which is a biodegradable plastic. One bacterium of interest for production of PHB is *Paraburkholderia sacchari* LMG 19450 LFM101. This non-model bacterium is a relatively high efficiency producer of PHB and can utilize both C5 and C6 sugars, making it an attractive candidate for PHB production from cheap feedstocks that may have a mixture of carbon sources, such as corn kernel fiber (CFK). In its native form, *P. sacchari* cannot efficiently produce enough PHB for the process to be economically feasible; however, there are limited synthetic biology tools available to attempt metabolic engineering strategies to rectify this shortcoming. Here we share our expansion on synthetic biology tools through characterization of a synthetic promoter set, development of an inducible system for gene control, and use of the RSF origin of replication in *P. sacchari*. We also expand on metabolic engineering strategies by demonstrating an increase in PHB production following expression of heterologous genes from another PHB producing bacterium, *Cupriavidus necator*. It is commonly recognized that transcriptional control is one of the easiest and most effective ways to control gene expression. To explore constitutive (always on) synthetic transcriptional control, we tested the Anderson promoter library for expression of GFP. We saw a range of expression which provides options for constitutive control of gene expression. We also sought to develop an inducible system within *P. sacchari*. The *lac* system from *E. coli* with the promoter P_{LacUV5} has been shown to achieve gene expression in *P. sacchari*, however, the off state is extremely leaky. We modified the *lac* system by changing the operator sites surrounding the promoter region to the Oid sequence shown to bind the repressor protein most strongly, resulting in a tighter off state when used to express green fluorescent protein (GFP). The presence of varying concentrations of the inducer molecule exhibited a graded expression response, therefore providing a tunable and useful tool for future metabolic engineering of *P. sacchari*. We also demonstrate that the RSF origin of replication is a candidate for plasmid maintenance within *P. sacchari*. Our findings show that a plasmid containing RSF and red fluorescent protein (RFP) as the reporter can express RFP within *P. sacchari*. This provides another origin of replication option, in addition to the previously used pBBR1-oriV, for plasmid- based gene expression within *P. sacchari*, which will be a useful tool for future multi-plasmid expression strategies. Finally, we have been able to demonstrate increased production of PHB from *P. sacchari* by incorporating heterologous genes from *C. necator*. A set of genes from *C. necator* (*phaA*, *phaB*, *phaC*, and *bktb*) were each individually inserted via plasmid and maintained with kanamycin resistance. These genes were chosen as they code for key enzymes in the PHB pathway which have been shown to work efficiently and we hypothesized that overexpression could increase PHB production. After 48 hours growth in nitrogen limited medium with 15 g/L glucose, there was a 3.5X increase in PHB production with the *P. sacchari* strain containing *bktb* under the constitutive control of the promoter P_{lac}. This gene, *bktb*, codes for β-ketothiolase, which converts Acetyl-CoA to Acetoacetyl-CoA as part of the PHB production pathway. Current and future work incorporates the synthetic biology tools described here into further exploration of how heterologous genes can improve PHB production within *P. sacchari*. Our findings lay a foundation for further development of a bioproduction system for PHB production; a direction of research that will provide a sustainable, environmentally responsible, and economically beneficial method for creating biodegradable plastics.

630359 Using Aqueous Renewable Solvents to Recover Lignin from Hybrid Poplar Lignin Cake

Abstract:

As biorefineries push closer to commercial viability, maximizing the usable yield of biomass is imperative for competing with traditional non-renewable alternatives. Lignin, as the only abundant aromatic biopolymer found in nature, can be used as a precursor in place of petrochemicals for applications such as carbon fibers, polyurethane foams, and activated carbon. Biomass from the hybrid poplar tree, a hardwood, has been targeted as a biorefinery feed crop, as it grows rapidly, is low maintenance, and has a high energy return. A by-product of enzymatic ethanol production from woody biomass, lignin cake, contains about 65% lignin, 35% sugars, and less than 1% ash by weight. To our knowledge, little research has been devoted to the purification of this lignin-rich waste stream. Thus, the proposed uses for lignin cake (aside from incineration for power generation) are scarce. Using renewable organic solvents with sulfuric acid as a catalyst, a high-purity lignin with very low ash and sugar content has been recovered from the impure lignin cake stream. The renewable solvents acetic acid and ethanol were used to dissolve the lignin, with the remaining solids being filtered out. Ash, glucose, and xylose content were measured via HPLC to determine the purity of the recovered lignin stream. Absolute (vs. relative by GPC) molecular weight was measured via SEC-MALS to determine the best application for the lignin fraction. Results for one of the applications delineated above will be presented.

628704 Crystallization of Phosphorus-Incorporated Solids from Liquid Phase of Hydrothermal Carbonization of Cow Manure

Abstract:

Crystallization of Phosphorus Incorporated Solids from Liquid Phase of Hydrothermal Carbonization of Cow Manure

Saeed V Qaramaleki, John A Villamil, Angel F Mohedano, Charles J Coronella

Excessive amount of manure produced in concentrated animal feeding operations (CAFOs) pose serious environmental and economic challenges. At the same time manure as biomass feedstock offers a huge potential as a resource of nutrients and energy. Therefore, developing processes to utilize manure as a resource can have substantial benefits. Hydrothermal carbonization (HTC) has been investigated to produce solid biofuels from various forms of biomass in recent years. As a thermochemical process, HTC can convert low energy value manure into a carbonaceous solid fuel primarily through various deoxygenation reactions. HTC can achieve this through a network of different reactions taking place in an aqueous environment which gives this process a big advantage over other techniques since it eliminates the need to dry the feedstock[1]. Operating at elevated but subcritical temperatures and pressures results in a dramatic change in physical properties of water. Subcritical water as reaction medium has a positive affect on the progression of hydrolysis, dehydration and decarboxylation reactions[2]. However, hydrochar is not the only product obtained from HTC, other byproducts including process water is produced as well. The focus of current study is on the liquid byproduct from HTC process. We have previously investigated the influence of various HTC conditions on solubilizing nutrients allowing nutrient recovery as a byproduct separate from the fuel value. In present work, we studied crystallization and precipitation of phosphorus-containing solids from HTC aqueous product, therefore taking one more step forward towards closing the phosphorus loop.

HTC of cow manure was carried out at 170°C in different acidic conditions, the concentration of acids was constant at 0.3 molar. The reaction time was 10 minutes for each run. The liquid product obtained by applying these conditions was then used as supernatant for crystallization. Due to presence of numerous kinds of inorganic and organic species, thermodynamic equilibrium of the solution is complex. Any change in the condition of the solution such as pH, chemical activity, or temperature, can move the solution into supersaturation. In these experiments, the solution pH of the hydrothermal aqueous product was increased by adding NaOH . Moreover, extra ammonia and magnesium was supplied in order to adjust the ratio of the N/P and Mg/P. All the precipitation experiments were conducted at room temperature and the pH of the solutions was fixed at 9.5. Nitrogen to phosphorus molar ratio was set to 2 by adding extra ammonium chloride. Three different Mg/P molar ratios (1, 1.5 and 2) were studied in order to determine the favorable condition for higher phosphorus recovery.

Our results indicate that precipitation of phosphorus-containing solids from HTC liquid is heavily dependent on the type of the acid used during HTC. As can be seen from the graph, adding either formic or sulfuric acid allowed for highest P recovery in the precipitate, relative to addition of other acids. The recovery of P is relatively insensitive to Mg:P for these two acids. The reason for a better recovery efficiency of these acids might be attributed to the smaller molecular weight of the acids used. For citric acid the recovery efficiency is almost constant with regards to the change in dosage of Mg (18.1-20.6%). Conversely, addition of oxalic acid indicates noticeable variability in terms of P recovery efficiency.

[1] Kruse A.,Funke A.,and Titirici M.-M., Hydrothermal conversion of biomass to fuels and energetic materials, *Curr. Opin. Chem. Biol.*, **2013**, 17, 3, 515–521.

[2] Funke A. and Ziegler F., Hydrothermal carbonization of biomass: A summary and discussion of chemical mechanisms for process engineering, *Biofuels, Bioprod. Biorefining*, **2010**, 4, 2, 160–177.

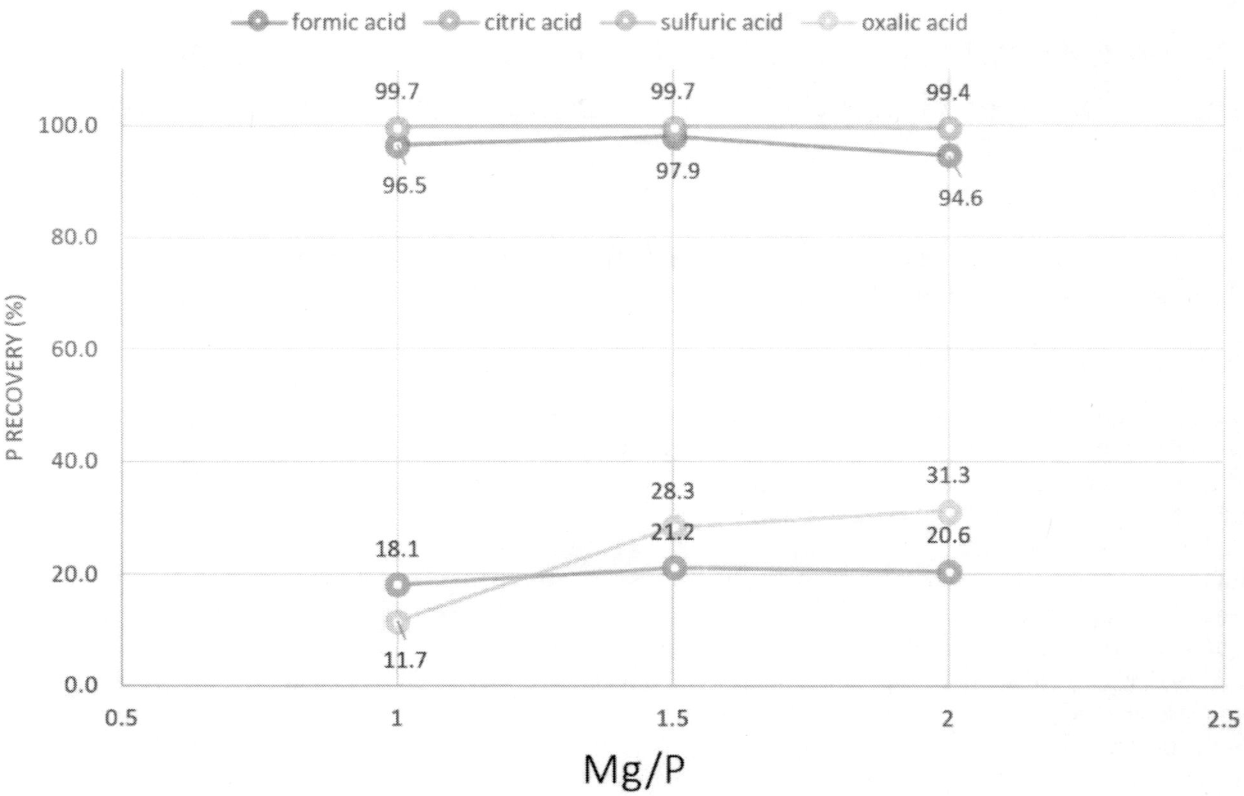
formic acid
citric acid
sulfuric acid
oxalic acid
P RECOVERY (%)
Mg/P
100.0
80.0
60.0
40.0
20.0
0.0
0.5
1
1.5
2
2.5
99.7
99.7
99.4
96.5
97.9
94.6
28.3
31.3
21.2
20.6
18.1
11.7

624996 Biomass Derived Sustainable Materials and High-Performance Devices

Abstract:

Plastics impregnate all aspects of our daily lives. Now white pollution plagues the planet. Creating a solid platform for growth of sustainable alternatives to fossil-based products is important. Biomass, as the largest biomaterial on the earth, is abundant, sustainable, and versatile. Cellulose, hemicellulose, and lignin are three primary earth abundant biomaterials derived from biomass; however, cellulose is mainly used for paper fabrication, and most lignin and hemicellulose are under-utilized. In this talk, we will introduce our studies on broadening the application of biodegradable materials from biomass with small environmental footprint, such as nanocellulose for bio/electronic/energy, tannin from bark for pseudocapacitor, lignin from wood as anolyte in flow batteries, galactomannan from seeds for transient biosensor, indigo carmine from flowers as anolyte in flow system, compressible carbon aerogel made from bacterial cellulose as anode of K ion batteries and so on. Meanwhile, due to the metabolism, wood tracheids are perfectly aligned for the transportation of water and ions. In this talk, we will also introduce our works on using the well aligned structure for mass and charge transportation in both energy storage and energy generation.

631616 Progress in Thermochemical Production of Cellulosic Sugars

Abstract:

Progress in Thermochemical Production of Cellulosic Sugars

Robert C. Brown

Iowa State University, Ames, IA 50011, U.S.A

rcbrown3@iastate.edu

AIChE Annual Meeting

Boston, MA

November 7-9, 2021

The conventional approach for production of cellulosic sugars from lignocellulosic biomass involves pretreatment to open up the cell wall structure followed by acid or enzymatic hydrolysis to convert polysaccharides into fermentable sugars. Enzymatic has been favored in recent years because of its selectivity toward sugars. Despite advances over the last several decades, commercialization of biofuels via enzymatic hydrolysis remains elusive. Among the major challenges has been the high cost of pretreatments to separate polysaccharides from lignin, production of cost-effective enzyme cocktails, and effective use or disposal of process wastes.

Research at Iowa State University is advancing thermal depolymerization of biomass, using heat and catalysts instead of enzymes to break glycosidic bonds in plant polysaccharides, as an alternative pathway to cellulosic sugars. This pathway has been relatively little explored, likely because the low selectivity and unfavorable thermodynamics of high temperature thermal processes would suggest poor prospects for thermal depolymerization. Indeed, pyrolysis of most lignocellulosic biomass yields very little sugar, with yields that are at best only a few percent of the theoretical yields of hexose and pentose from cellulose and hemicellulose. We have found that alkali and alkaline earth metals (AAEM) normally chelated into the polymeric structure of biomass can be ion exchanged with sulfuric acid or ferrous sulfate to form AAEM salts that are thermally stable at pyrolysis temperatures. The large anions in these salts shield the smaller AAEM cations from interacting with the biomass, thus effectively passivating their catalytic activity, and dramatically increasing sugar yields from the pyrolysis of biomass. We have learned how to recover these sugars from other products of pyrolysis and purify them for use in fermentations or chemical synthesis for production of various products. This paper examples the pretreatments, processing conditions, separations, and purifications required to produce pure monosaccharides via thermochemical processing.

637986 Single-Use Plastic Wastes: Challenges and Sustainable Alternatives

Abstract:

Plastic is an essential part of our daily lives, from everyday products to medical and transportation applications. Today, plastics is also a world's paramount environmental problem. Our inability to properly manage plastic disposal has resulted in persistent plastic waste with negative impact on the environment. Currently, the world produces around 450 million tons of plastics per year that is expected to double to around 1 billion tons per year, by 2050. Today, 50% of the plastic produced is for single-use only. The major share of plastics is used in packaging. Packaging, including single-use plastic, has a short life time; from limited days to fewer than six months. Thus managing its disposal is critical. When left in the environment, plastic does not biodegrade and instead breaks down to smaller parts including micro-plastics and has a devastating impact on ecosystems. The Ellen MacArthur Foundation's New Plastic Economy Initiative has set actionable targets to reach "100% reusable, recyclable or compostable plastic packaging by 2025". This presentation will focus on three categories of packaging structures that are most challenging to manage and therefore need new solutions. First, packaging with mixed materials, that combine plastic, paper and metals. Second, multi-layer protective films that recycling facilities cannot separate easily and therefore do not get recycled in general. Third, and most important, are items that are impractical to recycle even though they are mono-materials, like disposable cutlery, straws, and takeout containers. Many provinces and cities around the globe are banning single-use plastic products. Such a ban as a solution needs to be coupled with viable alternatives. Biodegradable plastics show promise as an alternative to certain types of single-use plastic. Innovative technologies are needed to enable a wider adoption of these new promising materials. In addition, the "end-of-use" for such new biodegradable and compostable plastic products should be adequately addressed. An integrated approach with regulatory policies, incentives, options for proper disposal, and other creative paths can address the complex journey of developing and implementing sustainable materials. These new materials can be used as an effective tool to alleviate environmental pollution caused by plastic wastes more importantly the single-use plastics.

Acknowledgements: (i) The Ontario Ministry of Agriculture, Food and Rural Affairs (OMAFRA) – University of Guelph, the Bioeconomy Industrial Uses Research Program Theme (Project Nos. 030361, 030486 and 030578); (ii) OMAFRA-University of Guelph Gryphon's Leading to the Accelerated Adoption of Innovative Research (LAAIR) Program (Project No. 030416); (iii) the Ontario Research Fund, Research Excellence Program; Round-9 (ORF-RE09) from the Ontario Ministry of Economic Development, Job Creation and Trade, Canada (Project No. 053970); and (iv) the Natural Sciences and Engineering Research Council of Canada (NSERC), Discovery Grants (Project No. 401111).

TABLE OF CONTENTS

628259 Could Benzyl Hydroxyl Shielding Promote the Radical Induced Pyrolysis of Lignin?

Abstract:

Introduction

Due to the inherent complexity and changing chemical structure of lignin, a strategy, which has been gradually accepted in liquid depolymerization, is to shield the benzyl hydroxyl to inhibit the self-condensation reaction of lignin. This information on whether benzyl hydroxyl shielded in lignin structure affected the pyrolysis behaviors of lignin, especially explained by the radical theory, however, is scarce in the literature. Accordingly, the influence of the shielded benzyl hydroxyl in lignin structure on thermal behavior and radical evolution was investigated in this study. The benzyl hydroxyl in the lignin structure was first stabilized by propionaldehyde and prepoxidized by 2,3-dichloro-5,6-dicyano-1,4-benzoquinone (DDQ) respectively, compared with the raw lignin as a control (ball-milling wood lignin, MWL) for subsequent investigation. Then, the raw lignin and pretreated lignin were pyrolyzed at different temperatures, and the corresponding products were determined to find the best condition for obtaining organic liquids. In addition, through the electron paramagnetic resonance (EPR) spectroscopy, the evolution of pyrolytic radicals had been comprehensively studied to reveal the connection of the products formation with the radical induced reaction by different method of pretreatment.

Materials and Methods

The maple wood was first prepared for subsequently extraction of different types of lignin. Milled wood lignin (MWL) were obtained from the maple wood via the ball-milling method. Propionaldehyde stabilized lignin (PAL) was obtained by extraction in dioxane: water solutions with HCl and the propionaldehyde. To prepare the DDQ oxidized lignin (DOL), the maple wood was firstly oxidized by the DDQ oxidation with O_2(g) as the co-oxidants. Then the DOL was obtained by the ball-milling method extracted from the oxidized wood. TGA, 2D HSQC NMR and Ultimate analysis of raw and pretreated lignin samples were conducted to investigate the characterization and thermal stability.

Analytical pyrolysis of MWL, DOL and PAL was conducted with pyrolysis gas chromatography/mass spectrometry (Py-GC/MS) at 350-650 °C. Lab-scale pyrolysis was performed on a fixed bed to harvest chars from lignin. The radical concentration of sample was detected on a Bruker EMX-10/12 spectrometer.

Results and Discussion

The results showed that the process of inhibiting the coking of lignin cannot be achieved by shielding the α-OH. The maximum total yields (11.10×10^9 area/mg) of MWL lignin was obtained at 450 °C compared with the oxidized lignin (3.71×10^9 area/mg of DOL, 5.02×10^9 area/mg of PAL), which indicated that the yields of bio-oil was inhibited by oxidized α-OH of the lignin.

To understand the pyrolysis radical induction, the radical evolution of MWL was first detected by EPR from 350 to 650 °C. The pyrolysis radical concentration increased rapidly from 26.82×10^{17} spin/g to 245.04×10^{17} spin/g at 550 °C, which means the lignin undergo an explosive fission reaction. For PAL radical detected at 550 °C, the lower radical concentration (169.32×10^{17} spin/g) and higher char yields (37.10%)was obtained compared to the MWL. As for DOL radical, the liquid yield of the DOL was lowered than the MWL with the higher spin concentration (259.32×10^{17} spin/g for DOL). Therefore, the possible pathway of radical evolution leading to various products distribution from raw and pretreated lignin was given a new sight to understand the lignin pyrolysis reaction.

Conclusions

The shielded α-OH in lignin structure showed significant effect on the thermal stability and yield distribution during fast pyrolysis. It was proved that the radical reactions play an important role and process of inhibiting the coking of lignin cannot be achieved by shielding the benzyl hydroxyl. The bio-oil yield was decreased after shielding the benzyl hydroxyl, while the char yield had been increased. Facts have proved that in the pyrolysis process, due to the violent radical-induced reaction, the caped hydroxyl groups might be quickly converted into oxygen radicals, which further increases the probability of formation of benzoquinone and caused carbonization. Thus, the effect of capping pretreatment on the distribution of lignin pyrolysis products is clarified, and the failure of benzyl hydroxyl shielding pretreatment and the possible inducement of promoting the condensation of lignin to carbon were proved from the perspective of radical induction.

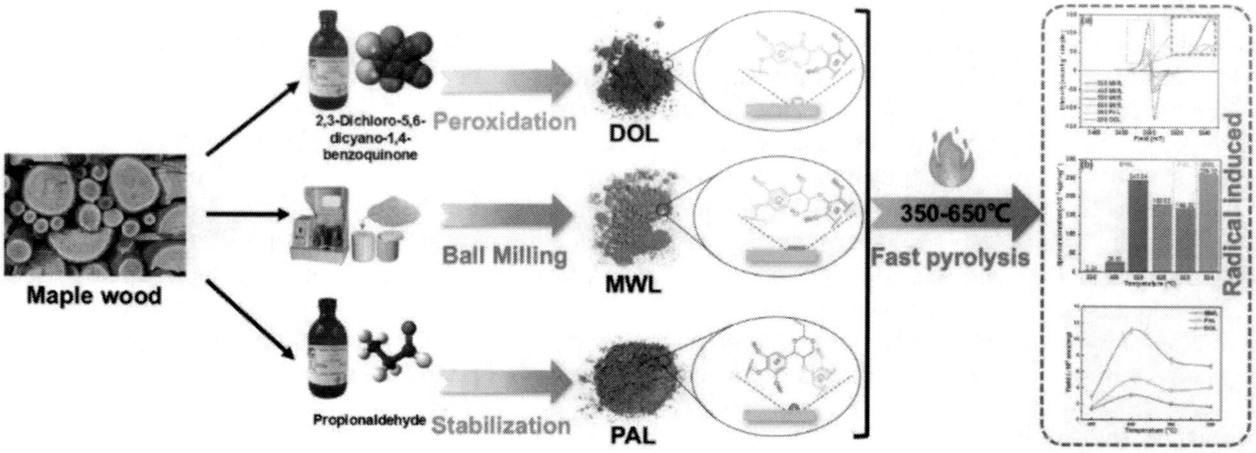

622248 Decarbonising the UK Heating Sector: A Whole-Energy System Analysis

Abstract:

The decarbonisation of heating is crucial in reaching the ambitious emission targets adopted by many countries, including the UK, which aims to reach net-zero emissions by 2050. While the electricity sector has received much attention from policymakers, private companies and researchers and considerable progress has been made on the decarbonisation of power generation, the heating sector has seen only little change. The majority of heat in the UK is still generated from conventional, unsustainable sources, with more than 85 % of households using natural gas boilers as main heating technology [1]. On top of that, 57 % of the commercial heat demand and 51 % of the industrial heat demand are provided from natural gas; with electricity, oil, solid fuel (coal) and biomass & waste also being significant contributors [2]. Consequently, emissions associated with heating are high. Residential combustion of fuel, mainly for heating purposes, alone accounted for 18 % of the UK's carbon emissions in 2019, and commercial and industrial heating add to that total [3]. It was estimated that in 2009, 38 % of the UK's carbon emissions could be attributed to heating [4].

Decarbonisation of heating is challenging due to the systems typically being distributed. Millions of devices in people's homes, commercial buildings and industrial facilities need to be replaced by low-emission heating technologies. These technologies typically rely on carbon-neutral fuels since carbon-capture at the end-user is impractical. Heat pumps and resistive electric heaters can utilise low-carbon electricity for heating, and hydrogen boilers can combust low-carbon hydrogen generated from e.g., methane-reforming with carbon-capture and storage (CCS) or water electrolysis without carbon emissions at the end-user. Solar-thermal collectors directly use a renewable energy source. Another potentially attractive option, especially for urban areas, are low-emission district heating networks that can benefit from centralised low-carbon generation and economies of scale. Additionally, improved building insulation and other energy conservation measures can be valuable as they decrease the demand and therefore reduce the required generation capacities. Various studies presented in literature show that no single technology is clearly superior to others, and most likely a combination of different technologies is most suitable for heating decarbonisation.

We present a newly developed heating sector optimisation model to identify optimal decarbonisation pathways. The model is a deterministic generation-expansion and unit-commitment optimisation model that simultaneously optimises technology portfolios and their operation. It is implemented as mixed-integer linear-programming (MILP) model in GAMS, using Python for data pre- and post-processing. A graphical user interface is also available to make the model accessible and easy to use. The model uses a similar formulation as the Electricity Systems Optimisation (ESO) model [5,6]. Starting from today's technology portfolio, the model optimises the evolution of the heating sector, considering investment decisions every five years. Build-rate constraints limit the deployment of technologies to technically feasible rates. For each of the decision years, the operation schedule of the technologies is optimised. Regarding emissions, the model can consider end-point constraints (e.g., net-zero in 2050), interior-path constraints and cumulative carbon budgets. So far, the model considers residential heating demand in form of representative households, commercial and industrial demand is to be added later. Modelled technologies are vapour-compression heat pumps, electric heaters, gas and hydrogen boilers and solar-thermal systems, and it is planned to include additional technologies such as thermal heat pumps, co-generation plants and district heating in future.

The emissions and fuel costs of electric and hydrogen heating technologies depend strongly on the respective generation mixes. A representation of the hydrogen industry is included in the heating sector model. Considered are generation from natural gas using methane reforming with or without CCS and from electricity via water electrolysis. Large-scale hydrogen storage is also included in the model. To determine the costs and emissions associated with electricity use, it is planned to connect the heating sector model with the existing ESO model. The result will be a comprehensive model of the UK electricity and heating energy systems that allows a simultaneous optimisation of technology capacities and dispatch in both sectors. The integrated approach allows to reduce the model externalities and necessary assumptions. The model can be a powerful tool to e.g., determine the optimal degree of heating electrification, prioritise the use of hydrogen for heating or electricity generation and to consider potential synergies such as co-generation or electricity load-shifting with heat pumps and thermal storage tanks. It can also be used to investigate the effectiveness of different policy measures. Guidelines on how to best support and facilitate the transition towards low-carbon heating and electricity provision can be developed. We present an analysis which robustly quantifies and qualifies the role and value of a range of technologies for providing low carbon heat.

References

[1] Committee on Climate Change, "Heat in UK buildings today" 2016. [Online]. Available: https://www.theccc.org.uk/wp-content/uploads/2017/01/Annex-2-Heat-in-UK-Buildings-Today-Committee-on-Climate-Change-October-2016.pdf.

[2] Department for Business Energy & Industrial Strategy, "Energy consumption in the UK (ECUK) - End use tables." 2019, [Online]. Available: https://www.gov.uk/government/statistics/energy-consumption-in-the-uk.

[3] Department for Business Energy & Industrial Strategy, "Final UK greenhouse gas emissions national statistics: 1990 to 2019." 2021, [Online]. Available: https://www.gov.uk/government/statistics/final-uk-greenhouse-gas-emissions-national-statistics-1990-to-2019.

[4] Department of Energy & Climate Change, "Overview of UK's greenhouse gas emissions from heat related activities" 2012. [Online]. Available: https://www.gov.uk/government/statistics/uk-emissions-from-heat.

[5] C. F. Heuberger, I. Staffell, N. Shah, and N. Mac Dowell, "A systems approach to quantifying the value of power generation and energy storage technologies in future electricity networks" Comput. Chem. Eng., vol. 107, pp. 247–256, Dec. 2017, doi: 10.1016/j.compchemeng.2017.05.012.

[6] C. F. Heuberger, E. S. Rubin, I. Staffell, N. Shah, and N. Mac Dowell, "Power capacity expansion planning considering endogenous technology cost learning" Appl. Energy, vol. 204, no. August, pp. 831–845, 2017, doi: 10.1016/j.apenergy.2017.07.075.

632205 Techno-Economic Assessment of Three Potential Pathways for Biomass Liquefaction

Abstract:

Biomass liquefaction is of interest within the biorefinery as it is a potential answer to the bottleneck formed from front-end materials handling. If the biomass could be transformed into a flowable slurry at the process entry point, then the material could more readily flow into the biorefinery pretreatment. Increased flowability will reduce downtime from any materials plugs that may form in the current design. Currently, the heterogeneous mixture of pelleted biomass and water can impede machine functionality and operability, which results in costly maintenance shutdowns. Liquefaction of biomass can potentially mitigate process bottlenecks. If a liquefaction process were to be installed at the front-end of an existing biorefinery, it would likely reduce or eliminate bottlenecks due to solids agglomeration. However, installation cannot be more costly than downtime losses. Through Techno-economic assessment (TEA), it is possible to quantify the reduction in losses via implementation of a liquefaction process. Three liquefaction pathways were considered: enzyme, maleic acid, and a combination of maleic acid+enzyme. The trade-offs in viscosity and equipment requirements were assessed using TEA to effectively compare the pathways to one another. All three pathways were simulated in Aspen Plus® Software.

623756 Altering Physico-Chemical Properties of Pine-Derived Carbon Quantum Dots By Changing Hydrothermal Treatment Conditions

Abstract:

Carbon quantum dots (CQD) are a relatively new and exciting material that are finding their place in a vast range of industries. These dots are aptly named as they exist within a diameter range of 0-10 nanometers. They are being utilized in everything from material coatings that self-clean or self-repair, to medicines that deliver cancer treatments to targeted areas in the body using UV light, to use in home entertainment systems to produce high quality screen visuals. Efficient methods and sources for producing these quantum dots are still being studied. It is found that the properties of the dots largely depend upon the surface functional groups and the size of the dots. Typically depending upon the method used to develop the dots (top-down or bottom-up) and the feed source the products are derived from will determine the surface groups and sizing.

A large amount of attention has recently been given to producing carbon quantum dots in the environmentally friendly manner of hydrothermal treatment. Many feedstocks have been studied and almost everyone has produced a variant dot. The goal of this research is to determine the effects of varying the hydrothermal treatment temperature and residence time on the characteristics of CQDs. Loblolly pine was chosen is chosen for this study due to abundancy and low cost. The experimental procedure focuses on three different reaction temperature conditions of 180, 220, and 260 °C. The time also varies in increments of 30 min up to 3 hours. After hydrothermal treatment the process fluid will undergo vacuum filtration followed by dialysis in order to separate the dots for characterization. Next the dots will be freeze dried for characterization that requires dry samples. The process fluid and the hydrochar itself will be analyzed for carbon quantum dots and these will be characterized using Fourier Transform Infra-Red Spectroscopy, Ultraviolet-Visible Spectroscopy, Fluorometry, Tunneling Electron Microscopy, and X-Ray Diffraction to determine trends. Altering the treatment time and temperature/pressure the quantity, surface functionalization, and the size can be controlled due to altering the formation of the passivation layer and condensation mechanism that forms the dots. As a result the quantum yield, absorbance, emission, and applicability of the dots are altered and can be fine tuned per application in industry.

623863 Understanding the Effects of Mixed Municipal Solid Waste Streams on HTL Product Characteristics

Abstract:

In 2018, each person in the United States was responsible for the production of nearly five pounds of municipal solid waste (MSW) per day.[1] Co-HTL of food and green waste components of MSW can help reduce waste and achieve economies of scale and the interactions between these feeds must be understood.

This work aims to understand the effect of mixing food and green waste on oil yield and product composition. Real food waste obtained from a veteran's affairs hospital resulted in 35.9% oil yield non-catalytically, in contrast to 12.13% for green waste. Current studies have revealed sustained oil yield (34.1%) with addition of up to 25% lignocellulosic waste (**Figure 1**). Additionally, results show increasing char yields (21 – 33.5%) with increased green waste content, due in part to its high (up to 13%) ash content.

Despite a drastic difference in carbon content in the feeds, 52% for food waste and 19% for green waste, FTIR indicates minimal differences between oil phase functional groups and vibrations. The primary difference occurs at 3400 cm^{-1} where the addition of green waste increases the OH stretching band. Solid phase FTIR mirrors the observed similarities, however, pure green waste solid phase which contains a significant OH stretch band. The similarities observed via FTIR are not sustained with GC/MS, as food waste oil contains large fatty acid peaks that decrease with the addition of green waste, only to be replaced by the presence of ketones and aromatics.

Qualitative observations also show improved homogeneity of food waste oil, noting that it is less sticky. Green waste oil, on the other hand appears to contain two distinct phases, a free-flowing liquid, and a sticky residue. This works provides useful insight into the feasibility of mixed MSW feeds for use in hydrothermal liquefaction.

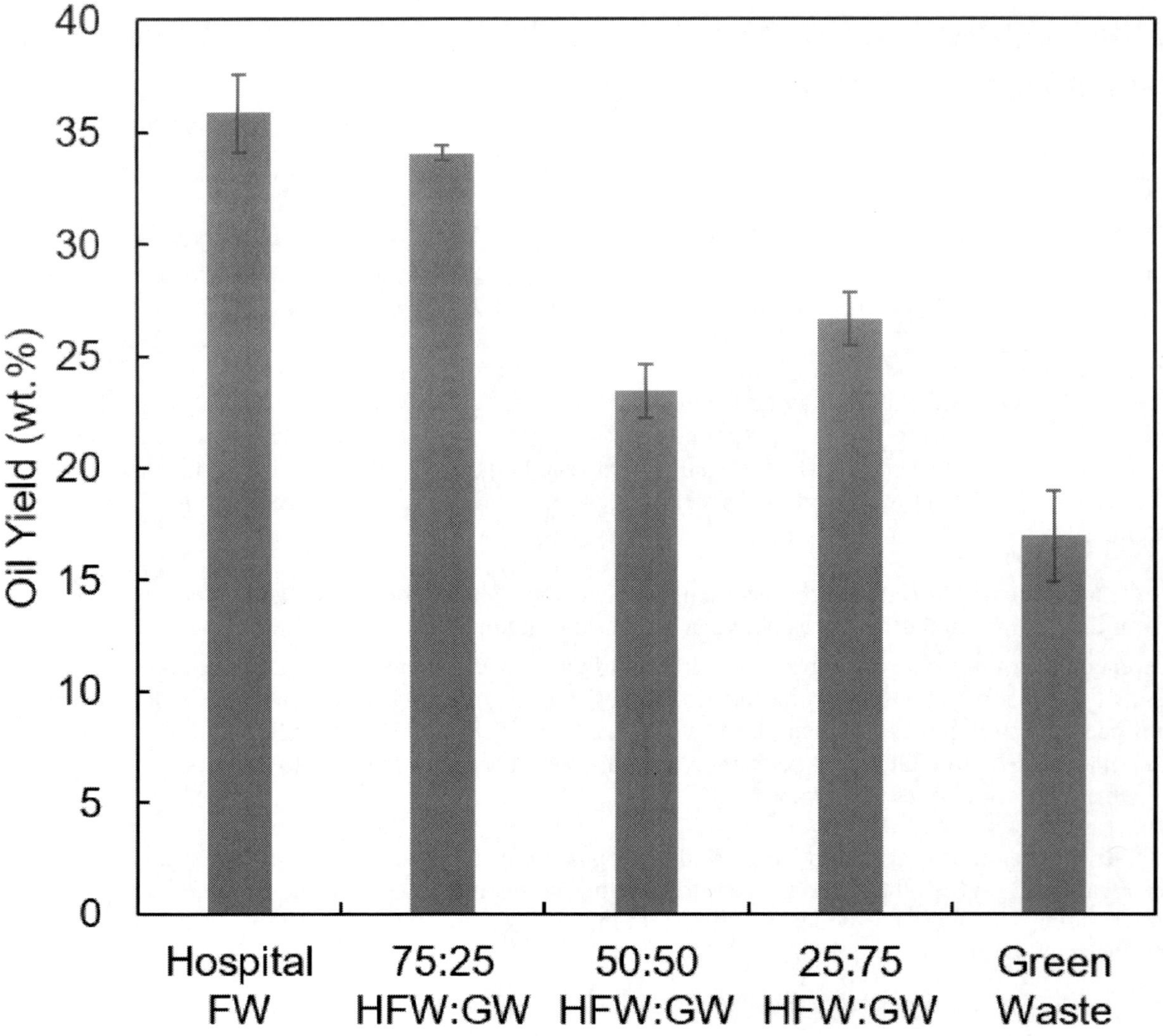
Oil Yield (wt.%)
0
5
10
15
20
25
30
35
40
Hospital FW
75:25 HFW:GW
50:50 HFW:GW
25:75 HFW:GW
Green Waste

632957 Adsorption Kinetics Studies of Catalytic HTL Derived Biochar

Abstract:

Ultra-porous carbonaceous powdered material derived from lignin rich biomass can be useful for water purification, energy storage, soil amendment, cement additive, wastewater treatment, gas purification etc.. In this study, we used corn stover agricultural residue as a cheaper feedstock and treated with hydrothermal liquefaction (HTL) at 250-300°C and 800-1600 psi to derive a solid residue, which was thermally and chemically post-treated to obtain mesoporous carbon. This carbon material was thoroughly characterized using the SEM/EDX, TEM and FTIR. Nitrogen adsorption isotherms such as Langmuir, Freundlich and Brunauer-Emmett-Teller (BET) were analyzed to determine the kinetic parameters. Barrett, Joyner, and Halenda (BJH) method was used to determine the pore-size distribution. Results obtained with specific surface area, pore size distribution, adsorption rate constant, monolayer/multilayer adsorption characteristics and mesoporosity will be presented.

629819 Cross-Flow Separation Characteristics and Piloting of Graphene Oxide Nanofiltration Membrane Sheets and Tubes for Kraft Black Liquor Concentration

Abstract:

Black liquor (BL) is a complex alkaline byproduct stream generated from the kraft process for biomass pulping at ~1 billion tons/yr globally. "Weak" BL from kraft pulping contains about 15 wt% dissolved solids including lignin, hemicellulose fragments, and inorganic salts. In kraft mills, BL is first concentrated/dewatered to 75-80 wt% solids and then combusted to generate steam and electricity while recovering the inorganic components as smelt for recycle. BL concentration by multi-effect evaporation is a highly energy-intensive operation. Membrane-based BL concentration up to 30 wt% solids is attractive for significantly reducing energy consumption, but conventional nanofiltration membranes are precluded by low solute rejections and poor stability in BL, which combines alkaline pH (~13) and high temperature (70-85°C). Our recent works [1,2] have demonstrated the engineering of graphene oxide (GO) membranes, their stability and promising performance in BL conditions, and preliminary scale-up into sheets and tubes. In this talk, we discuss in detail the separation characteristics of GO membrane sheets and tubes under real BL conditions and crossflow operation. The separation of lignin and hemicellulose fragments is characterized. Furthermore, detailed characterization of the permeation properties of the multicomponent inorganics (sulfate, sulfite, sulfide, thiosulfite, carbonate, hydroxide) is presented. We also present summary results of a slipstream pilot in a kraft mill. The results have strong implications for the industrial application of GO membranes in BL concentration and other related applications.

References:

1. Z. Wang, C. Ma, S. A. Sinquefield, M. L. Shofner, S. Nair, "High-Performance Graphene Oxide Nanofiltration Membranes for Black Liquor Concentration", *ACS Sustainable Chemistry & Engineering* 7 (17), 14915-14923 (2019).

2. Z. Wang, C. Ma, C. Xu, S. A. Sinquefield, M. L. Shofner, S. Nair, "Graphene oxide nanofiltration membranes for desalination under realistic conditions.", *Nature Sustainability* (2021). https://doi.org/10.1038/s41893-020-00674-3

622486 Purification of Lignin Monomers from Poplar Derived Reductive Catalytic Fractionation Oils with Counter-Current Chromatography

Abstract:

Reductive Catalytic Fractionation (RCF) is an emerging "lignin first" biomass pretreatment technique that produces a soluble stream of lignin compounds at appreciable concentrations. Direct isolation of these lignin compounds can yield high value co-products within a biorefinery. However, the isolation of individual compounds from RCF oil is challenging due to the inherent chemical complexity of the oil. For example, the use of traditional Simulated Moving Bed (SMB) technology for direct product isolation is problematic because dimers, trimers, and higher order partially depolymerized lignin present in the oil rapidly fouls the stationary phase. Thus, extensive upfront filtration, down to the nanofiltration level, is needed before SMB can be used. This filtration significantly lowers product yields (> 50%) and is not practical. One novel approach is to use Counter-Current Chromatography (CCC) to separate lignin monomers from RCF oil. CCC is promising process to recover bioproducts from these complex streams because it employs two liquid phases for both the mobile and stationary phase and therefore does not require a solid liquid (S/L) separation upfront. This is a form of process intensification, where the S/L separation occurs within the CCC device and is not required upfront as opposed to traditionally employed SMB. In this presentation, CCC will be discussed for separating bioproducts and biochemicals from RCF oils. Based on the experimental data, an economic analysis of the separation and recovery process will be presented and compared with processes using other state-of-art technologies including SMB.

625824 SWEET Sorghum Inbreeding to Enhance Sugar Control and Its Perspectives on Ethanol Production

Abstract:

Sweet sorghum is a promising energy crop for biofuel production due to its low input requirements, wide environmental adaptability, and high productivity (Mathur et al., 2017, Cheng et al., 2020). Sugars stored in sorghum stems are readily available for bioethanol conversion after harvest. Sucrose, fructose, and glucose are the three major sugars in sorghum juice (Kawahigashi et al., 2013). However, a wide range of Brix content is observed among different sorghum varieties. Genetic engineering and breeding technologies have been commonly applied to increase sugar content in sorghum stems (Bandara et al., 2020, Liu et al., 2021). In this study, the sugar concentration of sorghum juices collected from six sorghum inbreds and two seed set treatments were investigated. The sugar concentration ranging from one to 15 g/100ml was observed. The sorghum treated with bagging to prevent seed set resulted in a threefold higher sugar concentration than the sorghum allowed to set seed. Additionally, the high sugar tolerant ethanol red yeast was used to ferment sorghum sugar into ethanol. The rate of fermentation was fast, and all the sugars were completely depleted within 12 h, and the ethanol titers ranging from 59 to 65 ml/L were observed. The maximum theoretical ethanol yield of 95.87% was obtained for selected inbred. From the results, seed set treatments and sorghum genotype affected the juice sugar concentration. However, the theoretical ethanol yields were observed over 85%, regardless of various sugar concentrations derived from different sorghum inbreds and seed set treatments.

Reference:

Mathur, S., Umakanth, A.V., Tonapi, V.A., Sharma, R., Sharma, M.K., 2017. Sweet sorghum as biofuel feedstock: recent advances and available resources. Biotechnol. Biofuels. 10, 146.

Cheng, M.-H., Kadhum, H.J., Murthy, G.S., Dien, B.S., Singh, V., 2020. High solids loading biorefinery for the production of cellulosic sugars from bioenergy sorghum. Bioresour. Technol. 318, 124051.

Bandara, A.Y., Weerasooriya, D.K., Gobena, D.D., Hopper, D.J., Tesso, T.T., Little, C.R., 2020. Improving sweet sorghum for enhancing juice traits and biomass. Plant Breed. 139, 131-140.

Liu, G., Zhang Y., Gong, H., Li, S., Pan, Y., Davis, C., Jin, H-C., Wu, L., Godwin, I.D., 2021. Stem vacuole-targetted sucrose isomerase enhances sugar content in sorghum. Biotechnol. Biofuels. 14, 53

633258 Visualization and Characterization of Maize (corn stover) Cell Wall Deconstruction By Deacetylation, Ozonation, and Mechanical Refining Pretreatments

Abstract:

A novel pretreatment strategy for lignocellulosic biomass conversion called deacetylation and mechanical refining (DMR) improves sugar yields from enzymatic conversion, but the specific mechanisms involved in this technique are not well known. Visualization and characterization of pretreated corn stover was carried out to better understand how DMR improves biomass digestibility. Structural changes to the tissue structure were revealed by brightfield microscopy, where cells were separated and tissue structure was partially collapsed. Within the cell walls, delamination and an opening of the wall architecture was observed by 3D-electron tomography. In addition, microfibril separation perpendicular to the orientation of the cell wall lamella was shown. The direction of cell wall disruption is unique to DMR pretreatment and may aid in increasing sugar yields by providing more access for pretreatment chemicals and enzymes to enter the cell wall. Quantitative 3D structural analysis revealed large pores created in some biomass cell wall regions that could provide additional access for enzymes into the cell wall, even as some cell wall regions of DMR samples maintained characteristics similar to the control samples. This variable deconstruction provides an opportunity to improve DMR pretreatment to more uniformly deconstruct the cell wall.

633009 Development of a 3D Transport-Reaction MODEL to Understand the Pretreatment Processes in Plant Using RAMAN Spectroscopy

Abstract:

The renewable nature of plant biomass makes it an ideal raw material for the production of a wide variety of bio-based products that include pulp and paper, bioplastics, biofuels, wood plastic composites etc. In order to carry out biomass conversion to other products, the recalcitrant structure of biomass is first disrupted through a pretreatment process. This step also increases the efficiency of the subsequent hydrolysis step. Pretreatment can be carried out by mechanical, chemical, thermochemical or biological methods. During this process, one or more of the cell wall components are degraded and the cell wall structure is significantly altered. These changes in structure and topochemistry of the plant cell wall affect the transport and reaction during the pretreatment process. Hence, it is extremely important to develop a fundamental understanding of the underlying process in order to develop effective pretreatment methods.

In the current work, the structural and topochemical changes in lignin during the pretreatment of the biomass structure is obtained using Raman spectroscopy. In addition, a stochastic dynamic model using 3D structure of plant biomass is presented to evaluate the transport and reaction behavior of delignification during pretreatment processes. The lignin distribution map of the initial biomass structure obtained from Raman imaging is used as a starting point for this model. The transport-reaction model is based on a hybrid random walk process for diffusion of the pretreatment reagent followed by reaction with the cell wall components based on a certain reaction probability. The reagent walkers diffuse through the lumen and pore spaces of the cell wall and follow a random walk path until they encounter the cell wall interface. At the interface, based on the probability of reaction and the ratio of diffusivities between the pore space and cell wall, they either react with the lignin in the cell wall or diffuse further into the cell wall. The probability of reaction is based on a Thiele modulus which is a function of both diffusion and reaction. The degradation of lignin is modeled as a pseudo first order reaction where lignin is the limiting reactant. This process is continued with a sufficient number of reagent particles or walkers to simulate the entire course of the reaction and the results are compared with experimental data. This stochastic dynamic approach keeps track of the bulk concentration and spatial distribution of both lignin and the reagent used for pretreatment in real time. The results obtained from the model were compared with experimental data obtained from Raman spectroscopy. In addition, the variation of different diffusive and reactive parameters on the transport-reaction process were also studied. An effective rate constant K_{eff} based on the overall transport-reaction was determined for the process which was either time dependent or independent of time based on the time required for diffusion and reaction. Additionally, an overall transport rate coefficient K_T based on the survival time of the tracers was also determined.

Two cases of pretreatment corresponding to different types of biomass species and reagent were studied. These include alkali pretreatment of poplar and hot water pretreatment of Acer. The concentration profiles of lignin were obtained in each case and compared to the experimental data. The corresponding rate expressions were also determined. The results from the two cases were compared to study the effect of biomass species and the reagent on the pretreatment process. Thus, this model enables us to determine the effect of 3D structure as well as the effective diffusivity and local reaction kinetics on the overall pretreatment process and structural evolution. This will provide additional fundamental insights on biomass pretreatment and conversion process and develop effective biomass conversion strategies. Even more broadly, this can provide additional opportunities for better understanding the role of structure and reaction kinetics in porous materials such as catalyst pellets and their wide-ranging applications.

633557 Thiol-functionalized Hyper-Cross-Linked Milk Protein polymers for Mercury Removal

Abstract:

Hyper-cross-linked organic polymers (HCPs) have recently attracted a lot of attention as their large surface area and strong interactions with other molecules making them highly appropriate for different applications. The production of this type of microporous polymers is usually accomplished by employing pricey reagents. As an inexpensive way of synthesizing HCPs scaffold, we create one step Friedel–Crafts alkylation reaction with milk caseins, existing in high quantity and exhibiting unique polymer properties. To optimize the surface area of the polymer, the ratio of the catalyst ($AlCl_3$) to the cross-linker (2,4,6-Trichloro-1,3,5-triazine) was studied. Then, the polymer was decorated with sulfur functional groups to become a suitable platform for mercury separation from the water and gas phases. Our approach together with characterization techniques (FTIR, ICP, XPS, BET) validated that the integration of thiol-groups into HCP leads to remarkable de-mercuration performance. Thiol-functionalized casein polymer scaffold demonstrates excellent capacity and selectivity for trapping heavy metal ions. The high capabilities of developed sustainable adsorbents from waste streams make them comparable to current state-of-the-art nanoporous materials.

623747 Effect of Selective Impurities on Carbon Capture from Biogas Using Deep Eutectic Solvents

Abstract:

Anaerobic digestion (AD) of food wastes is one of the most promising pathways to convert wastes to value added fuels (e.g., biogas) and products (e.g., fertilizer). As food wastes are heterogeneous and their composition changes from household-to-household, the resulting biogas composition varies significantly. Biogas from food wastes often contain 40-60 % methane (CH_4), 40-60 % carbon dioxide (CO_2), and 1-5 % impurities. Although the actual concentration may vary from the source of biomass and AD operating conditions, most common impurities found in the biogas are hydrogen sulfide (H_2S), ammonia (NH_3), octamethyltrisiloxane, and decamethylpentasiloxane. carbon capture and removal of the impurities are required prior to use or injection into a natural gas pipeline. CO_2 from the gaseous stream can be separated by physical (ionic liquid) or chemical solvents (amine-based solvents). Amine-based chemical solvents have higher adsorption capacity but are often expensive to regenerate and toxic for the environment. Physisorption solvents are generally expensive and their adsorption capacity is lower, meaning that significantly higher amounts of physisorption solvents are required. In the context of carbon capture, the impurities are likely to play a significant role in adsorptive capacity of the solvents due to an influence on the molecular environment.

In this study, deep eutectic solvents (DES) are evaluated to upgrade biogas. The benefit of DES is that they are made from the physical mixing of a hydrogen bond donor (HBD) and a hydrogen bond acceptor (HBA), generally made from a quaternary salt and acid/alcohols. In this study, a conductor like screening model (COSMO) is adopted to screen suitable HBA's and HBD's for biogas upgrading based on hydrogen bond properties using a triple-zeta valence polarization (TZVP) basis set. This model is evaluated compared to literature experimental studies to verify integrity of the results. Of the five DES studied, it was found that quaternary-octyl ammonium bromide with decanoic acid ($N_{8888}Br$: DA) showed the highest CO_2 adsorptive capability. However, the choline-chloride (ChCl) based solvents had a much higher selectivity for CO_2 over CH_4. Through an activity coefficient analysis, it was determined that choline chloride-based DES are much more affected by the presence of contaminants than the ammonium salts with regards to adsorptive capabilities. It has been observed that the HBD is the main determinant of contaminant sensitivity rather than alkyl chain length. Due to these several trade-off factors apparent from the results, it is determined that the concentration of impurities should be accounted for when determining an appropriate DES for carbon capture from biogas.

629073 Design, Simulation, and Validation of 3-D Printed Hydrocyclones for Microbial Cell Concentration

Abstract:

Concentration of microbial cells is a key step in bioprocessing and laboratory operations. At the laboratory scale, concentration is typically conducted in centrifuges, however, this requires energy and consumables, and is not compatible with autonomous, continuous operation. Hydrocyclones are promising alternatives because the only energy input is to operate a pump, there are no consumables, the hydrocyclone requires no labor after setup, and it can operate continuously. This study provides a parametric open source design of a 3-D printable hydrocyclone with interchangeable parts for ease of testing different hydrocyclone dimensions during experimental optimization. The hydrocyclone was simulated using COMSOL Multiphysics® software using the k-ε turbulence physics to simulate rotational fluid flow and coupled with the particle tracing module to track individual cells or flocs. The simulation was able to replicate the rotational vortex and central air core and could track individual cells to the hydrocyclone overflow and underflow. The simulation was used to estimate the concentration efficiency for different hydrocyclone geometries. Simulations were experimentally validated for *E. coli* cultures mixed with varying concentrations of food-grade flocculant to increase the size of particles in solution and improve cell concentration efficiency.

629770 Switching Bioprocesses into Zero-Waste Virtual Power Plants By Flexible Operation of Electrochemical pH-Swing Extraction Processes

Abstract:

Carboxylic acids are important platform molecules that enable a sustainable chemistry based on renewable resources. [1] Although the biochemistry of the production is well established the successful industrial application is still challenging. As an example the US-Canadian company Bio-Amber, once considered a rising star for their ambition to establish a fermentation based production route for succinic acid had to file for bankruptcy. [2] This demonstrates that up to today advantages in environmental metrics cannot sufficiently compensate an economic disadvantage of the bio-based production. In order to make bio-based chemistry successful a holistic process development is required and must strive to achieve high yields in the fermentation steps, a low-cost and low-waste separation process and exploit all opportunities to harvest every added value feasible in the process.

We developed an electrochemically driven pH-swing extraction process for the recovery of succinic acid from aqueous solutions. [3] [4] This concept overcomes the issue of salt waste production otherwise encountered during the separation of bio-based succinic acid. [5] Instead of waste generating acid and base this technology uses electricity to drive the pH-swing extraction. The calculated electricity costs required to operate the electrochemical pH-swing extraction with the current prototype equipment are comparable to the costs for acid, base and waste disposal in state of the art process designs. A more detailed techno-economic assessment of the operational expenses (OPEX) of our electrochemical separation concept and four other process designs proposed in the literature was conducted with ASPENplus. The study revealed a cost advantage of the electrochemical processing route compared to other downstream options. Beyond the promising OPEX values determined for the electrochemical pH-swing process the technology enables utilization of electricity instead of fossil fuel derived heat for the separation of succinic acid.

Preferably the electricity used for the separation should be harvested from renewable sources like wind and solar power. These energy sources are inherently fluctuating and an extensive use of these sources will impose a significant stress on the stability of the power grid in the future. This talk demonstrates how a flexible operation of the electrochemical pH-swing extraction process can turn a bio-based succinic acid production into a virtual power plant. We implemented an algorithm in MATLAB that given the price profile of the electricity market calculates the cost-optimal load profile for the electrochemical unit operations. Modulating the power consumption according to the electricity prices provides a 7% saving on the electricity costs. A dynamic model of the electrochemical pH-swing extraction is used to calculate the electric current density, pump-speed and other operational parameters from the dynamic load profile. With the determined operational parameters the load flexible operation of a 100cm² electrolysis cell was tested experimentally. Throughout the experiment a constant degree of protonation close to 100% was measured, indicating that despite the load flexible operation a constant separation efficiency is viable with the operational conditions suggested with the model.

The developed method enables a load flexible and cost optimal operation of the electrochemical pH-swing extraction of succinic acid. In addition to the cost advantage harvested from times of abundant renewable electricity, hydrogen and oxygen are concurrently produced in the electrochemical pH-swing step. The cost advantage together with the generation of valuable co-products constitute an added value that can improve the overall economic competitiveness of the bio-based succinic acid production. Beyond that, load flexible operation and demand side management of large consumers aids in stabilizing the power grid and will likely become more important in the renewable centered electricity supply of the future.

Acknowledgements

The authors gratefully acknowledge the financial support of the Kopernikus project SynErgie (03SFK3L1-2) by the Federal Ministry of Education and Research (BMBF) and the project supervision by the project management organization Projektträger Jülich (PtJ).

Literature Cited

1. Becker J, Lange A, Fabarius J, Wittmann C. Top value platform chemicals: Bio-based production of organic acids. *Current Opinion in Biotechnology*. 2015;36:168–175.
2. Michael McCoy. Succinic acid maker BioAmber is bankrupt. *Chemical & Engineering News*. 2018.
3. AiChE Proceedings, editor. *An Electric Drive for Biorefineries: Electrochemical pH-Shift Liquid-Liquid Extraction of Succinic Acid.*, 2019.
4. Gausmann M, Kocks C, Doeker M, Eggert A, Maßmann T, Jupke A. Recovery of succinic acid by integrated multi-phase electrochemical pH-shift extraction and crystallization. *Separation and Purification Technology*. 2020;240(240):116489.
5. Kiss AA, Lange J-P, Schuur B, Brilman DWF, van der Ham AGJ, Kersten SRA. Separation technology–Making a difference in biorefineries. *Biomass and Bioenergy*. 2016;95:296–309.

632720 Electrokinetic Modeling of Salt Transport and Rejection in Graphene Oxide Nanofiltration Membranes

Abstract:

Graphene oxide (GO)-based membranes have gained tremendous interest in nanofiltration (NF) and reverse osmosis (RO) applications. A major challenge in GO membrane science is to improve their salt rejection in NF and RO processes. Previous researchers have proved that the interlayer spacing between the GO layers as well as the charge density of the GO layers can strongly influence salt rejections from ideal/dilute solutions. However, many practical desalination and dewatering processes operate at much higher concentrations. In this work, our goal is to systematically understand and predict the effect of salts concentration on GO membrane properties by electrokinetic modeling supported by experimental measurements. The correlations between feed concentration, interlayer spacing, and charge density of these GO membranes were determined by single-component salt permeation experiments and ion adsorption models. The modified Donnan steric pore model was then used to model, fit, and predict the salts rejections at different conditions. We then use the transport model investigate in detail the contributions of convection, diffusion, and electro-migration as a function of feed conditions. Finally, the combined effects of initial charge density and interlayer distances on salts rejection are evaluated through the plotting of 2D heat maps. Our study indicates that for neutral pH, under low salt concentrations (~ 0.01 M), modifying the charge density of GO will be more helpful for increasing salt rejections. On the other hand, under high concentrations (above 0.1 M), decreasing the interlayer distance of the GO membranes (i.e., reduction of effective pore size) would be a better strategy. Finally, we show the accurate prediction of binary/multicomponent salt mixture permeation behavior using the model. Overall, this study provides mechanistic insight into salt transport and rejection in GO membranes, and useful guidance for modification of GO membranes in a number of desalination processes.

622690 Process Design and Scale-up Study for the Production of Polyol-Based Biopolymers from Sawdust

Abstract:

Polymers are materials with exceptional properties that contributed to the development of modern societies (Thompson et al., 2009; Yarsley and Couzens, 1945). However, multiple environmental threats associated with their extensive use have been revealed. The difficult and slow degradability of these materials, which leads the generation of micro-plastics polluting soils and fresh and marine waterbodies, is an environmental issue of major concern. This requires the development of new environmentally and economically sustainable materials with adequate properties capable to replace traditional petroleum-based polymers. Different wastes are emerging as interesting raw materials for this purpose, which use to produce new polymers also contribute to reduce the environmental footprint of modern societies. Particularly, sawdust is a by-product of the forestry industry which can be used as raw material for the production of added-value bio-based chemicals, leading to the development of a circular economy based on this organic resource. One of the most promising paths for the reuse of sawdust is the production of polyols, which in turn can be used as raw material for obtaining biopolymers capable of replacing crude derived polymers, changing the paradigm of the polymer industry towards a more sustainable model.

A conceptual design and techno-economic evaluation of a process for the production of polyol-based biopolymers from sawdust at full-scale have been performed in this work. The process has been designed following the premise of being as environmentally sustainable as possible, but using simple equipment feasible to be deployed and operated by the timber facilities. It consists of preparing the raw material, the catalysis and glycerol for the solvolysis reaction followed by the purification of the polyol and finally the production of the bio polymer adding starch. A fraction of the sawdust processed is combusted to cover the thermal energy requirements of the process rather than using external fossil fuel supplies. The performance of the process stages is assessed based on experimental data (CIPA, 2017).

The results show that, as a result of the low capital expenses required, the deployment of the process proposed has a low economic entry barrier. However, the intensive use of raw materials results in most of the production costs are due to raw materials cost. In order to evaluate the effect of the scale in the economy of the process, a scale-up study has been performed. It is revealed that production costs are largely affected by the economies of scale, allowing a significant reduction of the investment per unit of waste processed as the facilities increase their size. For the typical range of market prices of glycerol and starch, the production cost ranges from 0.10 to 0.66 USD per kilogram of biopolymer produced.

References

CIPA, 2017. Biodegradable polymer based materials using Wood residues of SME from wood industry in Chile's Biobío region. R15F10009. Technical Report. Centro de Investigación de Polímeros Avanzados.

Thompson, R.C., Moore, C.J., Vom Saal, F.S., Swan, S.H., 2009. Plastics, the environment and human health: current consensus and future trends. Philosophical Transactions of the Royal Society B: Biological Sciences 364, 2153-2166.

Yarsley, V., Couzens, E., 1945. Plastics Middlesex. Penguin Books Limited.

624814 Phasins Employed By *Rhodopseudomonas Palustris* CGA009 for Bioplastic Production from Lignocellulosic Biomass

Abstract:

Lignocellulosic biomass is considered to be the most economic carbon source in the world, and is thus an ideal candidate for cheaper production of bioplastics. *Rhodopseudomonas palustris* CGA009 is a very metabolically diverse microbe that can utilize lignin breakdown products to produce bioplastics. Our recent efforts suggest cytoplasmic space is a limiting factor of maximum production of the bioplastic polyhydroxybutyrate (PHB) by *R. palustris*. Phasins are the dominant proteins surrounding PHB granules that control the size, shape, and abundance of granules. In this study, the Phap1 phasin from *Cupriavidus necator* H16 was expressed in *R. palustris* with the aim of overproducing PHB from lignocellulosic sources. Phap1 leads to smaller and more abundant PHB granules in *C. necator*, and it was hypothesized that expression of this phasin would increase overall titers from *R. palustris*. Transmission electron microscopy revealed that indeed Phap1 expression yielded smaller and more abundant granules in both aerobic and anaerobic conditions compared to the wild type strain. Expression of Phap1 yielded PHB production from *R. palustris* aerobically (0.7 g/L), which does not occur in the wild type strain. Although the Phap1 expression modified the granules in anaerobic conditions, the PHB titer was not as high compared to the wild type strain. This suggests that phasin performance may be optimal under certain conditions since *C. necator* is typically an aerobic microbe. Our ongoing efforts are aimed at quantifying changes in gene expression to develop a more holistic analysis of *R. palustris'* PHB production framework, including Phap1 as well as native phasins employed by *R. palustris* that have never been reported.

628379 Engineering Functional Materials from Cellulose Nanocrystals By Exploring Their Structure and Property Relationships

Abstract:

Cellulose nanocrystal (CNC) films possess chiral nematic structure by virtue of its chiral nematic self-assembly and therefore exhibits iridescence or structural color. Although, the structural color has been exploited for various application, it also restricts CNC's applications in packaging or coating industry where transparency is a prime factor.

The goal of this study is to evaluate the combined effect of different electrolyte addition and mechanical force alignment on CNC structure and properties. We found that particle order parameter of CNC increased from 0.32 to as high as 0.90 under the combined effect of electrolyte addition and mechanical force at a very low speed of 50 mm/sec. The scope of this study also includes comprehensive analysis of CNC structure and properties such as crystallinity, mechanical, optical, electrical, rheological, and barrier properties.

629832 Discovery and Development of New Sustainable Polyesters from Biomass

Abstract:

One of the greatest challenges facing the world today is the accumulation of plastic waste in the environment. Researchers have long pursued the development of biodegradable plastics, and some are already widely used, for example the popular poly(lactic acid) PLA or poly(hydroxybutyrate) PHB. However, the poor thermo-mechanical properties of such polymers limit their use for many applications. Therefore, it has become essential to discover and develop a new biodegradable polymer platform with robust mechanical properties, which can be derived from biosources to meet societal needs. In this work, we have explored the development of such new polyesters based on hydroxyacid monomers that can be produced from biomass. After searching the literature for a chemically diverse set of target molecules, this class of monomers was carefully evaluated for its polymerizability to narrow down to the most promising candidates. Initial efforts to synthesize high molecular weight homopolymer was challenging due to high crystallinity of short alkyl chain hydroxyacid compounds. However, copolymer made out of 3-hydroxypivalic acid and 6-hydroxyhexanoic acid exhibited high molecular weight as well as good thermo-mechanical properties that allowed the thermal processing window to be systematically widened by tuning copolymer composition. Lastly, a high throughput degradation was introduced and performed to assess the biodegradation of such polymers. The results from biological discovery, chemical synthesis and environmental metabolic engineering indicate that polyesters made from biosourced hydroxyacids have potentials to simultaneously meet the need for polymers with improved performance and improved sustainability.

631151 Modeling Heat Transfer and Reaction Kinetics of Biomass in Pyrolysis Feeding Systems

Abstract:

Biomass feedstocks can be used to produce low carbon-intensity chemicals and liquid fuels by displacing fossil-fuel sources. Pyrolysis is the thermal decomposition of biomass in the absence of oxygen and can be used to produce crude biofuels. A typical feeding mechanism for pyrolysis reactors is by lock hopper followed by a horizontal auger feeder. Particle agglomeration and plugging of the auger screw are regularly occurring phenomena that is detrimental to the process since it results in the need for periodic inspection and cleaning. The underlying mechanisms leading to agglomeration and plugging are unclear, but a working hypothesis is that heat from the reactor raises the temperature of the biomass in the feeder to the point at which preliminary decomposition reactions occur and produce "sticky" products. Using principles of heat and mass transfer, we derived one-dimensional differential equations for heat flow from the pyrolysis reactor through the auger feeder. Temperature profiles were solved from the system of one-dimensional differential equations using the shooting method. Subsequently, we used the temperature profiles as a one-way coupling to a reaction model for the biomass in the feeder. Kinetic models were adopted from the literature and used to predict the formation of intermediates, products, char, extractives, and metaplastics as well as phase changes along the auger feeder. Species were identified that correlated with softening of the biomass, causing it to become sticky and plug the feeder. We will discuss the details of this work, our findings, and the direction of future work that includes fully coupled transport and reaction-kinetics simulations.

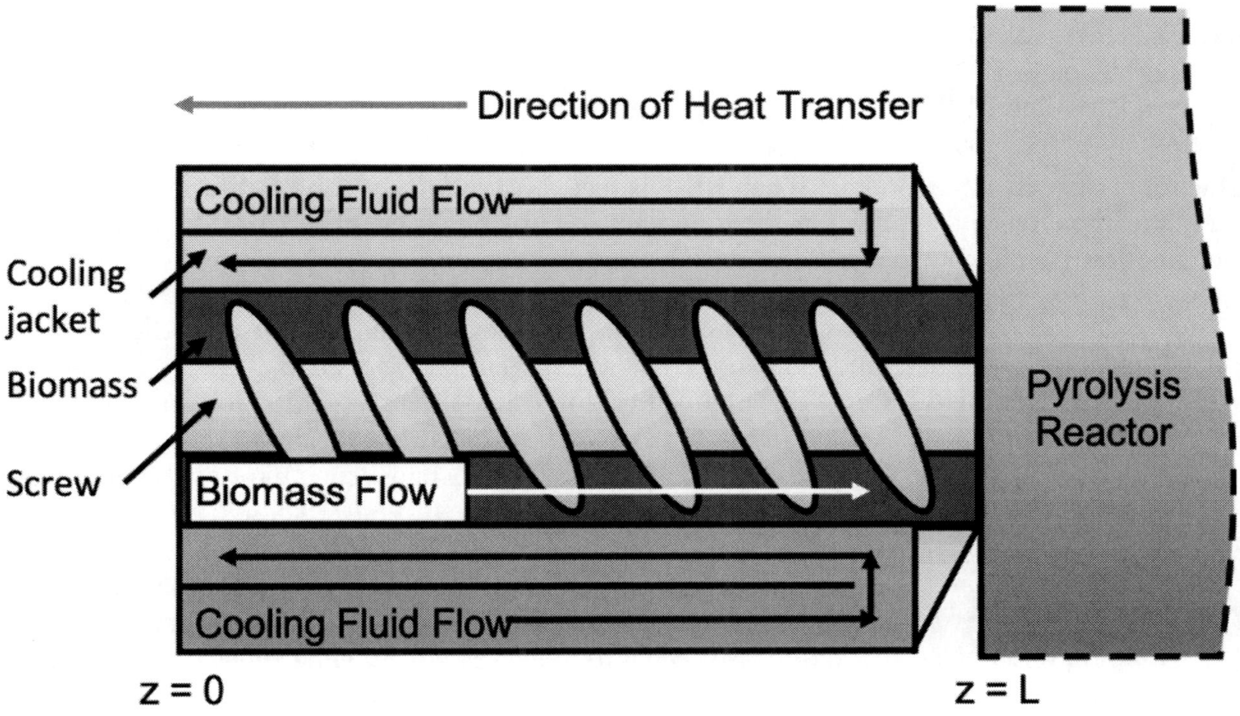

632423 High Yield Synthesis of HMF from Glucose in the Water-Organic Solvent System

Abstract:

5-hydroxymethylfurfural (HMF) is a versatile bio-derived platform molecule obtained dehydration of glucose or fructose. HMF selectivity and yield are higher when fructose is the starting material rather than glucose, but glucose is naturally-abundant in lignocellulose whereas resources of natural fructose are limited. Leveraging our Simultaneous-Isomerization-and-Reactive-Extraction (SIRE) followed by Back-Extraction (BE) process,[1-3] we present an integrated pathway for efficient conversion of glucose into HMF via fructose. In SIRE, aqueous glucose is enzymatically isomerized to fructose and simultaneously extracted into an immiscible octanol phase containing naphthalene-2-boronic acid (N2B) in a biphasic system. Due to the selective binding of N2B to fructose, the octanol phase has a high extraction selectivity for fructose (80-90%). Next, the fructose-rich octanol phase is contacted with acidic water, again in a biphasic system, to back-extract (BE) the sugars from the immiscible octanol phase. The acidic fructose-rich aqueous solution obtained from SIRE-BE is mixed with the co-solvent - 1,2-dimethoxyethane (DME) and heated to allow sugar dehydration and formation of HMF. After optimizing reaction parameters in the water-DME system, we report that with a concentrated sugar feed of 25 wt% (on water-basis, glucose: fructose = 1:4) with a HCl catalyst concentration of 55 mM (water-basis), high HMF yield of 87% (mol HMF formed/mol sugar reacted) can be achieved after dehydration at 120 °C for 30 min. Fructose is completely converted during this reaction, while glucose remains largely unreacted (glucose conversion = 18% (mol sugar reacted/mol sugar added)) . Recovery of HMF from the water-DME system is energy-efficient due to the low heat of vaporization of DME (383 kJ/kg, 85 °C). We also demonstrate a pathway to separate the unreacted glucose, HMF and DME by phase splitting using toluene as the ternary solvent. Our techno-economic assessment of the process estimates a minimum selling price of isolated HMF to be $1.22/kg when glucose is sourced at $0.236/kg.[4] When HMF is recovered in toluene for 2,5-diformylfuran and 2,5-furan dicarboxylic acid application, the minimum selling price is estimated to be $1.19/kg-HMF (toluene-free basis), which is best reported so far in the literature.

1. Li, B.; Relue, P.; Varanasi, S., Simultaneous isomerization and reactive extraction of biomass sugars for high yield production of ketose sugars. *Green Chem.* **2012,** *14* (9), 2436-2444.
2. Li, B.; Varanasi, S.; Relue, P., High yield aldose-ketose transformation for isolation and facile conversion of biomass sugar to furan. *Green Chem.* **2013,** *15* (8), 2149-2157.
3. Gogar, R.; Viamajala, S.; Relue, P. A.; Varanasi, S., Techno-Economic Assessment of Mixed-Furan Production from Diverse Biomass Hydrolysates. *ACS Sustainable Chemistry & Engineering* **2021,** *9* (9), 3428-3438.
4. Davis, R.; Tao, L.; Scarlata, C.; Tan, E. C. D.; Ross, J.; Lukas, J.; Sexton, D. *Process Design and Economics for the Conversion of Lignocellulosic Biomass to Hydrocarbons: Dilute-Acid and Enzymatic Deconstruction of Biomass to Sugars and Catalytic Conversion of Sugars to Hydrocarbons*; NREL/TP-5100-62498; National Renewable Energy Lab. (NREL), Golden, CO (United States): 2015; p Medium: ED; Size: 133 p

631345 Characterization of Transgenic Sugarcane (lipidcane 1566) and Its Potential As a Raw Material for Co-Production of Ethanol and Biodiesel

Abstract:

Biofuels and bio-products are green alternatives to the petro- fuels and -chemicals. Bioethanol and biodiesel are two major biofuels used in the transportation sector. The last decade has seen a 44% increase in fuel ethanol production to 15.8 billion gallons (Energy, 2021). Bioenergy crops like energycane, sugarcane, miscanthus and, sorghum have shown immense potential for biofuel production over recent years. These bioenergy crops primarily produce fermentable sugars which are further bio-processed to bioethanol. Plant oils have high energy density and are catalytically upgraded to biodiesel and a wide range of products, including chemicals (Lligadas et al., 2010; Salimon et al., 2012).

Considering the growing plant oil market to make biodiesel, challenging agricultural land use and high production costs make biodiesel production from current oil crops (soybean) unfeasible (Huang et al., 2016; Huang et al., 2017). Accumulation of TAG in the vegetative tissues such as leaves and stems of bioenergy grasses discounts major drawbacks of traditional oil resources (Long et al., 2015). TAG functions as a storage lipid in plants and recent achievements show that plants can be genetically modified to increase TAG content (James et al., 2010). Lignocellulosic biomass with high triacylglyceride (TAG) content and fatty acids rich in short, unbranched, and unsaturated side chains are suitable candidates for biodiesel production. Interestingly, to this end, breakthrough research is being done to genetically modify the metabolic pathway of bioenergy crops to enhance the accumulation of oilseed-like energy-rich TAG molecules in their vegetative cells. Investigators have successfully reported an increase in TAG accumulation as high as 20 and 25-fold in *Nicotiana tabacum* and *Arabidopsis thaliana*, respectively, on constitutive co-expression of a cluster of genes responsible for TAG accumulation and reduction of lipid turnover (Andrianov et al., 2010; Sanjaya et al., 2013). This proof of concept was challenging to apply in high biomass C4 perennial grasses. C4 plants have high photosynthetic efficiency and hence, have more capacity to convert solar energy into chemical energy in the form of storage lipids. However, various research groups are working to apply the above-mentioned concept in different bioenergy crops i.e., sugarcane, energycane, miscanthus, and sorghum.

Recently, Zale et al (2016) reported a 1.5-9.5 fold increase in TAG accumulation in vegetative tissues of sugarcane (Zale et al., 2016). The lipid producing lines of sugarcane are designated as lipidcane. Total lipid content of 0.9-1.3% was extracted from the juice and bagasse of the lipidcane lines. An analysis of the lipid composition showed that TAG constituted approximately 31 to 33% of total lipid and then investigated the feasibility of utilizing plant oils as a platform chemical to make biobased products (Huang et al., 2017). The economic analysis of biodiesel and ethanol co-production from lipid-cane showed that transgenic lipid-cane can produce 6700 liters of biodiesel from a hectare of land while soybeans can produce 500 liters. TAG being the platform compound for the production of high-quality biodiesel, lipidcane certainly exhibited potential as a cellulosic feedstock for biodiesel production. However, sugarcane is a seasonal crop and it cannot be grown around the year.

In this work, we characterized lipidcane (1566) grown in Illinois (IL) and Florida (FL) and compared it to the wildtype sugarcane grown under the same conditions. The composition of stem and leaves were determined using the National Renewable Energy Laboratory's analytical procedure – *NREL/*

TP-510-42618. The stem and leaves of IL-grown lipidcane contained 16.87% and 27.37% extractives respectively. The stems of FL lipidcane contained approximately 10% more extractives and the leaves had nearly 13% less, compared to IL lipidcane leaves. The IL lipidcane leaves recorded the lowest glucan concentration at 25.80%, while FL lipidcane leaves contained 32.61% glucan. The lipidcane stems of FL and IL contained 29.72% and 33.03% glucan and xylan concentration at 17.49% and 16.78% respectively.

For biodiesel production, the composition of TAG and free fatty acids plays a deciding role in the final quality (Knothe, 2009). Thus, biomass pretreatment needs chemical-free and low severity physical methods to prevent the degeneration of oil. Lipidcane and wildtype were pretreated at 180°C for 10 min, using a chemical-free hydrothermal pretreatment technique. The pretreated sample was disk milled and then enzymatically hydrolyzed to yield monomeric sugars. Subsamples from fresh stem and leaves, pretreated and enzymatically hydrolyzed material were centrifuged. The supernatant and solids were freeze-dried and then analyzed for fatty acids and TAG distribution. The lipid content was quantified by gas chromatographic (GC) analysis of fatty acid methyl esters (FAME). The gas chromatographic retention times relative to those of FAMEs of known structures were used to identify FAMEs. The identities were verified by gas chromatography-mass spectrometry. The free fatty acids and TAG got concentrated in the pretreated biomass and the solids remaining after hydrolysis. For instance, pretreated Illinois grown lipidcane contained 35.94 μg/mg dry weight (DW), which concentrated to 47.53 μg/mg DW in the leftover solids after hydrolysis. The stems from the same plant contained 35% less concentrated free fatty acids. TAG was present at 10.65 μg/mg in pretreated leaves and 3.21 μg/mg in pretreated stems of Illinois grown lipidcane. FL lipidcane contained 8.14 μg/mg of fatty acids and 3.12 μg/mg of TAG in leaves and 5.19 μg/mg of fatty acids and 4.23 μg/mg of TAG in stems.

The composition of cellulosic biomass along with the accumulation of TAG molecules is a pleiotropic effect, hence is expected to be controlled by different growth conditions. Total lipid and sugar recovery from transgenic Lipidcane 1566 grown in Illinois and Florida will be compared to investigate the favorable growth conditions that support the maximum accumulation of energy-rich TAG molecules. The data from enzymatic hydrolysis, composition, and lipid analysis will be combined, reflecting a broad overview of sugar, lipid, and other components (lignin, extractives, ash) distribution in a FL and IL lipidcane plant. Total biodiesel and bioethanol production per hectare of IL and FL lipidcane cultivation will be discussed. The total hectares of lipidcane cultivation necessary to replace soybean biodiesel obtained per hectare of agricultural land will be addressed using the lipid yield of IL and FL lipidcane. Finally, the improvements to *in-situ cellulosic* lipid production will be proposed to substitute a hectare of soybean cultivation and produce same quantity of biodiesel.

References

Andrianov, V., Borisjuk, N., Pogrebnyak, N., Brinker, A., Dixon, J., Spitsin, S., Flynn, J., Matyszczuk, P., Andryszak, K., Laurelli, M., Golovkin, M., Koprowski, H., 2010. Tobacco as a production platform for biofuel: overexpression of Arabidopsis DGAT and LEC2 genes increases accumulation and shifts the composition of lipids in green biomass. Plant Biotechnology Journal 8(3), 277-287.

Energy, E.T., 2021. U.S. Energy Information Administration (EIA) Monthly Energy Review. https://www.eia.gov/totalenergy/data/monthly/#renewable. (Accessed 02/24/2021 2021).

Huang, H., Long, S., Singh, V., 2016. Techno-economic analysis of biodiesel and ethanol co-production from lipid-producing sugarcane. Biofuels, Bioproducts and Biorefining 10(3), 299-315.

Huang, H., Moreau, R.A., Powell, M.J., Wang, Z., Kannan, B., Altpeter, F., Grennan, A.K., Long, S.P., Singh, V., 2017. Evaluation of the quantity and composition of sugars and lipid in the juice and bagasse of lipid producing sugarcane. Biocatal Agric Biote 10, 148-155.

James, C.N., Horn, P.J., Case, C.R., Gidda, S.K., Zhang, D.Y., Mullen, R.T., Dyer, J.M., Anderson, R.G.W., Chapman, K.D., 2010. Disruption of the Arabidopsis CGI-58 homologue produces Chanarin-Dorfman-like lipid droplet accumulation in plants. P Natl Acad Sci USA 107(41), 17833-17838.

Knothe, G., 2009. Improving biodiesel fuel properties by modifying fatty ester composition. Energ Environ Sci 2(7), 759-766.

Lligadas, G., Ronda, J.C., Galià, M., CáDiz, V., 2010. Plant Oils as Platform Chemicals for Polyurethane Synthesis: Current State-of-the-Art. Biomacromolecules 11(11), 2825-2835.

Long, S.P., Karp, A., Buckeridge, M., Davis, S., Jaiswal, D., Moore, P., Moose, S., Murphy, D., Onwona-Agyeman, S., Vonshakh, A., 2015. Feedstocks for biofuels and bioenergy. Souza GM, Victoria R, Joly C et al.., et al.., eds. Bioenergy & Sustainability: Bridging the Gaps. Paris: Scientific Committee on Problems of the Environment (SCOPE), 302-347.

Salimon, J., Salih, N., Yousif, E., 2012. Industrial development and applications of plant oils and their biobased oleochemicals. Arabian Journal of Chemistry 5(2), 135-145.

Sanjaya, Miller, R., Durrett, T.P., Kosma, D.K., Lydic, T.A., Muthan, B., Koo, A.J.K., Bukhman, Y.V., Reid, G.E., Howe, G.A., Ohlrogge, J., Benning, C., 2013. Altered Lipid Composition and Enhanced Nutritional Value of <em>Arabidopsis</em> Leaves following Introduction of an Algal Diacylglycerol Acyltransferase 2. The Plant Cell 25(2), 677-693.

Zale, J., Jung, J.H., Kim, J.Y., Pathak, B., Karan, R., Liu, H., Chen, X., Wu, H., Candreva, J., Zhai, Z., Shanklin, J., Altpeter, F., 2016. Metabolic engineering of sugarcane to accumulate energy-dense triacylglycerols in vegetative biomass. Plant Biotechnology Journal 14(2), 661-669.

[+] Correspondence to: Vijay Singh, Department of Agricultural and Biological Engineering, University of Illinois at Urbana-Champaign, Urbana, IL, 360 G Agricultural Engineering Science Building 1304 W. Pennsylvania Urbana, IL 61801. E-mail: vsingh@illinois.edu

625646 Hot-Water Pretreatment and Saccharification of Genetically Modified Cellulosic Feedstock for Fuel Production

Abstract:

Demand for transportation fuel, bioethanol and biodiesel, is ever increasing and they are significant contributor to a country's economy. Increased research has been directed towards finding alternative renewable sources for the production of bioethanol and biodiesel. Understanding plant lipid biosynthesis and advances in genetic engineering tools has enabled researchers to develop modified crops capable of lipid production and accumulation. Lipids induced in vegetative tissue of energy crops are being projected as an alternative source of oil for biodiesel production. Beneficially, such crops are aimed to be cultivated on marginal land, thereby reducing stress on vegetable oil and avoiding the "food vs fuel" debate. Moreover, as cellulosic energy crops are used as a substrate for metabolic engineering to induce lipid, the cellulosic pretreatment approaches could be applicable for sugar and lipid recovery.

In this work, hot-water pretreatment to deconstruct the lipid containing cellulosic biomass to concentrate the lipid in residue and enable efficient enzymatic hydrolysis of polysaccharide backbone was evaluated. The biomass (bagasse) was composed of 35.32 ± 1.85% glucan, 20.62 ± 2.46% xylan and 17.01 ± 0.97% acid insoluble lignin. The biomass was pretreated at 170, 180, 190°C for different reaction times at 10% w/v solid loading. It was observed that under the optimal pretreatment condition, the cellulose content of the residue increased due to solubilization of hemicellulose and other components (Table 1). The residues and the autohydrolysate lipid content were quantified (Table 2) and analyzed for the triglyceride content (Figure 1). It can be noted that a concentration of triglycerides in the residue occurs upon pretreatment and the highest concentration was observed at 170°C. The pretreatment condition where highest lipid and cellulose content was observed in residue was selected for subsequent enzymatic hydrolysis. The pretreated residue was enzymatically hydrolyzed into constituent monosaccharides using a mixture of cellulases and hemicellulase. The residues and the liquor after enzymatic hydrolysis were analyzed for lipid content. The preliminary results are indicative of an increase in lipid content upon pretreatment and enzymatic hydrolysis. Moreover, high conversion of polysaccharides into constituent monosaccharides (glucose and xylose) was observed.

Thus, in this work, a cellulosic pretreatment approach was used to pretreat lipid-producing genetically engineered crop to obtain lipid and sugar. Thus, a single feedstock could potentially enable the production of two renewable fuels, biodiesel and bioethanol.

Table 1: Composition of cellulosic residue after hot-water pretreatment

Temperature (°C)	Time (min)	Log Ro	Glucan (%)	Xylan (%)	Acid Insoluble Lignin (%)
170	10	3.061	55.64 ± 2.07	11.59 ± 0.25	23.81 ± 0.22
170	20	3.362	56.09 ± 3.44	4.78 ± 0.23	31.99 ± 2.14
180	10	3.355	59.73 ± 3.52	8.70 ± 1.28	25.23 ± 0.36
180	20	3.657	59.48 ± 1.31	4.04 ± 0.07	29.94 ± 0.39
190	5	3.349	55.23 ± 2.71	13.54 ± 0.62	22.85 ± 0.10
190	10	3.650	59.88 ± 1.85	6.02 ± 0.20	27.19 ± 1.26

Table 2: Lipid content of bagasse and pretreated samples

Time	Temperature	Log Ro	% Lipid (Residue)	% Lipid (Liquor)
biomass			6.3 ± 0.11	
10	170	3.061	6.86 ± 0.07	0.58 ± 0.19
20	170	3.362	6.82 ± 0.37	0.52 ± 0.19
10	180	3.355	4.78 ± 0.1	1.05 ± 0.05
20	180	3.657	3.32 ± 0.56	0.85 ± 0.81
5	190	3.349	4.19 ± 0.21	1.11 ± 0.07
10	190	3.650	3.53 ± 0.65	1.72 ± 0.67

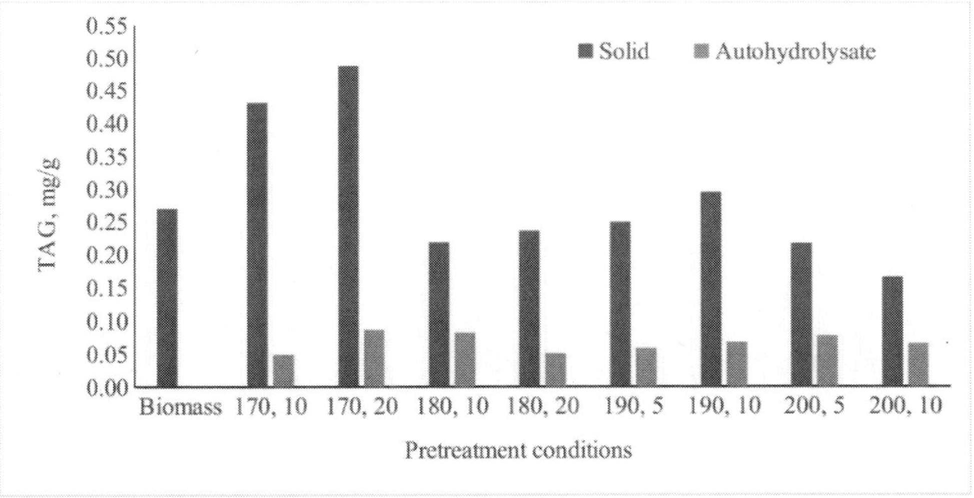

Figure 1: Triglyceride content of biomass and pretreated residue and autohydrolysate. The first number indicates temperature (°C) and the second number indicates time (minutes)

626572 Coupled Near Infrared Spectroscopy and Air Classification of Corn Stover for Improved Feedstock Quality

Abstract:

A principal challenge of both lignocellulosic and petroleum-based fuel production is heterogeneity of feedstock materials. While both biomass and crude oil require separation to produce more homogeneous fractions, the latter has centuries of technical development. To be competitive, front-end biomass separation must be scalable and inexpensive: constraints met by pneumatic separation (air classification). Air classification leverages differences in density and aerodynamic profiles of biomass anatomical tissues to homogenize the feedstock. Near infrared spectroscopy (NIRS) enables high throughput composition analysis that can be correlated to anatomical tissue type. Here, we demonstrate that a physics-based pneumatic separation model enhanced with NIRS can inform front-end separation processes to improve feedstock quality. Our work shows that NIRS predicts anatomical tissue composition of corn stover while calculated partition velocities predict observed air classified corn stover fractions. Coupling these independently powerful tools advances separation technology to a state that can be leveraged by industry to enable cost-competitive, at-scale biofuels.

623760 Lignin Extraction from Cellulose in Loblolly Pine Using Deep Eutectic Solvent Screening with COSMO

Abstract:

Lignin and cellulose are the two most abundant natural polymers with wood being the main global source. Cellulose is a sugar polymer which can be converted into biofuel. Lignin is a complex organic structure that encapsulates the cellulose. Due to the vast abundance of these materials and their propensity for being discarded as waste by industry, there exists a strong argument for using it as a cheap and sustainable source of biofuel. The process in which cellulose is converted to biofuel consists of two main stages, hydrolysis, and fermentation. The combined structure of lignin and cellulose undergo hydrolysis to break the sugar polymers of cellulose into glucose monomers. The glucose is then fermented into an ethanol and water mixture which can be separated to obtain the value product of ethanol. During these two processes the lignin disrupts the reaction times and, in some cases, poisons the process. This causes a lower process efficiency and drives up costs making the marketability of the product less feasible.

In this study, deep eutectic solvents (DES) are evaluated to extract lignin. The benefit of DES is that they are made from the physical mixing of a hydrogen bond donor (HBD) and a hydrogen bond acceptor (HBA), generally made from a quaternary salt and acid/alcohols. In this study, a conductor like screening model (COSMO) is adopted to screen suitable HBA's and HBD's for Lignin extraction based on hydrogen bond properties using a triple-zeta valence polarization (TZVP) basis set. This model is evaluated compared to literature experimental studies to verify integrity of the results. Several tetra alkyl ammonium HBA were combined with HBD's for this study. HBA were also combined with mixtures of HBD to search for synergistic effects that may be present. While no increases in performance were observed, decreases in performance were also absent. This suggests that using a mixed waste stream of acids to form DES is a potential strategy for industry. It was found that choline based HBA had the greatest capacity for solvating the lignin while leaving the delicate cellulose intact. This is likely due to the cleaving of non-hydrolyzed sections of the lignin driven by the higher polarity of the choline chloride-based solvents.

626800 One-Step Extraction of Multifunctionalized Lignins from Biomass

Abstract:

Lignin, one of the main components of biomass, is a promising feedstock to replace aromatic materials that are currently sourced from non-renewable fossil oil. However, the industrial use of lignin in material application remains limited[1]. This can partially be explained by the uncontrolled side-reactions that happen on the lignin during traditional extraction methods, which despite providing lignin at high yields, irreversibly change its structure. The poor control on the chemical functionalities is what ultimately hinders lignin's further upgrading[2]. We have recently demonstrated that with the Aldehyde Assisted Fractionation (AAF) it is possible to overcome the trade-off between lignin's isolation yields and structure preservation[3]. Moreover, when multifunctional aldehydes are employed, lignin can also be functionalized in a controlled manner[4].

Here, we show how this concept can be expanded to simultaneously extract and introduce more than one non-native functional groups on the lignin backbone. We functionalized the lignin with aldehydes by using terephthalaldehyde (TALD), and carboxylic acids by using glyoxylic acid (GA). First, we isolated several lignins by performing AAF with increasing ratios of TALD and GA, and then we quantified the introduced functional groups with several spectroscopic techniques. The results clearly showed that the TALD/GA ratio initially used for the AAF linearly correlated with the TALD/GA ratio measured on the isolated lignin.

The different functionalities of TALD/GA lignins were then employed to improve the mechanical properties of gelatin-based hydrogels. This lignin with two distinct and controlled non-native functionalities can in fact be used to impart multiple features to the lignin. For example, the carboxylate groups imparted water solubility to the lignin, while the aldehyde functionalities acted as a chemical crosslinker in the gelatin-lignin adduct.

Overall, we demonstrated that multifunctional lignins can be designed according to their final application and subsequently extracted in a controlled way and in high yields in one single step from lignocellulosic biomass. The ultimate goal will be to integrate these lignins into novel materials for high-end applications.

References:

1 V. Romhányi, D. Kun and B. Pukánszky, *ACS Sust. Chem. Eng.*, 2018, **6**, 14323–14331.

2 S. Bertella and J. S. Luterbacher, *Trends in Chemistry*, 2020, **2**, 440–453.

3 L. Shuai, M. T. Amiri, Y. M. Questell-Santiago, F. Héroguel, Y. Li, H. Kim, R. Meilan, C. Chapple, J. Ralph and J. S. Luterbacher, *Science*, 2016, **354**, 329–333.

4 S. Bertella and J. S. Luterbacher, *Green Chem.*, 2021, In revision.

629747 Deashing of Biomass By Deep Eutectic Solvent to Enhance Biomass Conversion Process

Abstract:

High ash biomass contains significant amount structural inorganics (monovalent, divalent and polyvalent metals) which affect the overall biomass quality as a biofuel. Hydrothermal carbonization (HTC) is a conventional and emerging thermochemical processing technology, where biomass is converted to low-ash, carbon-rich, energy-dense hydrochar. Yet, one of the shortcomings of HTC is that the polyvalent metals are entrapped in the complex structural matrix and almost impenetrable for leach out. Deep Eutectic Solvent (DES) is such a unique solvent that can leach structural inorganics out of biomass during pre-HTC treatment and augment in effective mass transfer during HTC. The purpose of this study is to pretreat raw biomass (e.g. Corn Stover High Ash Fraction, obtained from air classification technology) with Type-III DESs, i.e. choline chloride and urea (1:2 molar ratio), methyltriphenylphosphonium bromide and ethylene glycol (1:4 molar ratio) and observe the material, chemical and morphological change. The DESs, above their eutectic point, are mixed with the raw biomass and shaken for 30, 60, 90, and 120 mins at 155 °C in the oil bath. The treated biomasses are vacuum filtered with DI water and confirmed of no residual DES left by conductivity test. Finally, the biomasses are hydrothermally carbonized at three different temperatures (180 °C, 220 °C, 260 °C), characterized through thermogravimetric analysis, scanning electron microscopy (SEM), SEM-energy dispersive x-ray (SEM-EDX) and induced coupled plasma-optical emission spectrometry (ICP-OES, and compared with untreated hydrochars of the same temperatures. It is hypothesized that the DES interacts with the polyvalent metals through softening the lignin-(hemi)-cellulose (LCC) complex as found in Van Soest fiber analysis method and aids in deashing (ash reduction). Overall, the pretreatment could notably improve the biomass fuel property and can be regarded as an important preprocessing step for higher biomass conversion process.

630608 Lignin-Based Deep Eutectic Solvent Pretreatment of Transgenic Sweet Sorghum Bagasse to Achieve a Sustainable Biorefinery Process

Abstract:

Conversion efficiency is one of technical and economic obstacles in lignocellulosic biomass utilization. In this study, a closed-loop process with lignin-derived deep eutectic solvent (DES) pretreatment and genetically modified *p*-Hydroxybenzoic acid (PB)-rich sweet sorghum bagasse was investigated. DES synthesized with choline chloride (ChCl) and PB has shown high pretreatment efficiency. Results show that ChCl-PB DES effectively fractionated both wild-type and transgenic sweet sorghum bagasse at mild temperatures. The cellulase accessibility was significantly improved after the pretreatment. Formation of aqueous DES with a small quantity of water enhanced its pretreatment performance further. Recovered lignin was characterized with two-dimensional heteronuclear single quantum coherence (2D HSQC) nuclear magnetic resonance (NMR), ^{31}P NMR, and gel permeation chromatography (GPC) to understand the interaction between DES and biomass. The pretreatment preserved about half of the amount of β-O-4 linkage with relatively uniform sizes and low molecular weight. No condensed lignin unit was observed. This study indicates that the integration of lignin-based DES with genetic engineering is a promising strategy to develop a sustainable biorefinery process.

629729 Resolving the Discrepancies in the True Molecular Weight of Lignins with the Assistance of the ALPHA Process

Abstract:

Lignin constitutes nearly 30% of all renewable carbons in the biosphere and hold tremendous potential value as a source material for a wide breadth of downstream applications. However, the complexity and heterogeneity of lignin streams, as a result of the delignification process and natural variety in source materials, make it exceptionally difficult to develop. Advancement of lignin technologies hinges on the ability to clean and purify technical lignins into well-defined and fractionated streams. Here we employ a patented process titled "*Aqueous Lignin Purification with Hot Agents*" (ALPHA). The characterization of complex polymers is a bottleneck for sustainable material development, with lignin being no exception. Currently, there are enormous discrepancies in literature regarding the true molecular weights of lignin and presents a notable obstacle for improving purification, fractionation, and material models. One promising technique for calculating the absolute molecular weight and distribution of lignins is size-exclusion chromatography (SEC) coupled with multi-angle light scattering equipped with an infrared laser ($MALS_{IR}$) and bandwidth filters. This approach both suppresses and blocks fluorescence, respectively, which has led to overestimation of molecular weight of previous MALS experiments. We demonstrate how SEC-$MALS_{IR}$ can be used to accurately calculate the molecular weight distributions of a variety of lignin sources and ALPHA-processed fractions. A universal-calibration style technique is applied to compensate for lingering non-SEC interactions, fluorescent, and absorption related phenomena that cause erroneous and poorly reproducible measurements. In doing so, we decouple discrepancies discovered between mobile phases, lignin fractions and the individual contributions of non-ideal behavior during liquid chromatography and light scattering. Moving forward, this work presents an improved methodology of using a SEC-MAL_{IR} technique to help bridge the characterization gap that inhibits meaningful lignin development across the field.

633223 Integrated Biochemical and Hydrothermal Processing of Corn Stover for Fuels and High Value Products

Abstract:

Biochemical processing involving alkaline/acidic pretreatment followed by enzymatic hydrolysis of corn stover generate sugars as well as unhydrolyzed solids (UHS). Traditional fermentation route can be followed to convert sugars into alcohols whereas UHS can be treated with hydrothermal processing for high value products such as hydrochar and bio-oil. In this study, both catalytic and non-catalytic hydrothermal processing of preprocessed corn stover was performed for the production of fuels and high value products. Hydrochar obtained was selectively treated via thermal/catalytic route to convert into high surface area mesoporous graphitic carbon for charge storage application. In addition, hydrothermally derived hydrochar was converted into another high value product such as carbon nanofibers. Bio-oil was made into blend fuels using diesel and ethanol. ASPEN modeling was performed to scale-up the process at 1 tpd. A low-value stream generated in terms of the aqueous waste from the hydrothermal processing of UHS was enriched with lactic acid and phenol. Effect of high value products on fuel cost reduction was attempted. Results obtained on energy and mass balance, products yields and quality. and techno-economic analysis will be presented.

622236 Synergistic Effects on Co-Hydrothermal Liquefaction of High Ash Corn Stover and Halogenated Plastic

Abstract:

Agricultural residues can be thermally processed through a variety of methods to produce bio-oil or bio-crude that can be upgraded to chemical and drop-in biofuels or even combined in a traditional refinery. Corn Stover is considered as one of the key lignocellulosic feedstocks for biofuels production. One promising method of biomass conversion is hydrothermal liquefaction (HTL), where lignocellulose residues are broken down in water at high temperature and pressure to produce a bio-crude oil. Additionally, polyvinyl chloride (PVC) is the major halogen source in municipal solid waste and its recycling is problematic. Hydrothermal treatment has been proposed to be an effective method to dehalogenation of PVC, since hydrochloric acid is solubilized in subcritical water and act as a catalyst to enhance the biomass degradation as well as solubilizing inorganics from biomass. Therefore, this study aims to investigate the co-hydrothermal liquefaction (co-HTL) of PVC with high-ash corn stover in subcritical water (280-350 °C). Our intention was to derive a paramount benefit from the catalytic effect of HCl on the yield instead of using additional catalysts. In this work, the effects of blending ratio of PVC/CS (0.05 – 0.25) and reaction time (30 and 60 min) on the biocrude yield and quality are evaluated through ultimate analysis, proximate analysis, gas chromatography mass spectroscopy, thermogravimetric analysis and Fourier-transform infrared spectroscopy. Our results indicate that synergistic effects can be observed by co-HTL in terms of biocrude yield and quality in compared to HTL of standalone feedstocks.

627387 Catalytic Hydropyrolysis of Rice Husk over a Hierarchical Micro-Mesoporous Composite Catalyst

Abstract:

Catalytic fast pyrolysis (CFP) has been widely used in research to convert biomass into liquid hydrocarbons. It consists of thermal decomposition of the organic molecules from biomass in inert atmosphere and in the presence of a heterogeneous catalyst. Unfortunately, there are critical problems in this traditional thermal cracking method, such as catalyst coking and low carbon efficiency, since large fractions of carbon initially present in the biomass end up in the coke and carbon-containing gases, resulting in a liquid phase product with relatively low hydrocarbons content. In contrast, catalyst hydropyrolysis (CFP) uses hydrogen as the reaction medium, which potentially may provide higher yields of aromatic hydrocarbons.

This work presents a comparison between the products from catalytic fast pyrolysis (CFP) and catalytic hydropyrolysis (CHP) of rice husk over a hierarchical catalyst. The results suggest that the addition of hydrogen inhibits polycondensation reactions, leading to a 2.3 wt. % reduction in coke yield at 400 °C and 35 bar. In addition, the presence of the hydrogen at these conditions increases the relative content of hydrocarbons in both liquid (by 16 %) and gas (by 2.7 %) products, suggesting that hydrodeoxygenation, decarbonylation, and decarboxylation reactions are promoted. The gaseous hydrocarbons promoted in the presence of hydrogen include CH_4 and C_2-C_3. In addition, we report the effect of catalyst characteristics on the products from catalytic hydropyrolysis of rice husk. Alkali-treatment of the HZSM-5 improves the catalytic activity and coking resistance of catalyst. The addition of an extra external MCM-41 layer to the alkali-treated HZSM-5 further improves the catalytic activity, resulting in a relative hydrocarbon content of 48.9%, which is higher than that obtained with alkali-treated HZSM-5 (36.8%) and pristine HZSM-5 (28.6%). We also report the effect of pressure on the products. As the pressure increases from 1 to 35 bar, the relative content of hydrocarbons increases from 39.9 % to 48.9 %, coke yield decreases from 3.4 wt. % to 1.9 wt. %, and CH_4 and C_2-C_3 yields decrease from 10.3 wt. % and 5.1 wt. % to 14.2 wt. % and 6.7 wt. %, respectively. As the temperature increased from 300 to 400°C, the relative content of hydrocarbons increased from 38.2% to 48.9%. Both CH_4 and C_2-C_3 yields reach the highest values of 14.2 wt. % and 6.7 wt. % at 400 °C. The results from this work highlight the effect of the hydrogen atmosphere for the thermal decomposition of rice husk over the hierarchical catalyst.

631573 Effect of Recycling Potassium Hydroxide on Surface Morphology Superactivated Hydrochar Derived from Loblolly Pine

Abstract:

Potassium hydroxide (KOH) is an effective activating agent for the development of substantial surface area and pore volume in activated carbons where more than 200 publications are reported annually (Scopus), most of which utilize KOH for the purpose of chemical activation. Consequently, a considerable fraction of KOH is released into the aqueous environment during the washing process of the activated carbon that promotes eco-toxicity in water. While almost none of the published studies emphasizes on the environmental effect of the released effluent containing residual KOH after activation, our research tapped on the potential of recycling the KOH after subsequent activation so that volume of KOH containing water discharged could be reduced. A novel technique was adopted in our study where the pH of the mixture that constituted the carbon precursor prior to activation (loblolly pine derived hydrochar) and the activating agent solution were every time maintained at pH = 13.5. Summarizing the process, firstly the hydrochar was soaked in the KOH solution of 3.6 mol/l (20 ml) for 2 h at room temperature which was then ultrasonicated for 6 h at room temperature. Afterwards, the solution was evaporated at 80°C overnight (16 h) and the residue was activated at 800 °C for 2 h. Following the activation, the sample was washed with 20 ml of water under vacuum filtration system while the pH of the filtrate was measured repeatedly to determine the volume of KOH solution of 3.6 mol/l required to be added to the filtrate until the pH reached 14. The resulting solution was used for its successive activation cycle following the same process of KOH impregnation as described above for a total of 10 cycles. Surface morphology of the activated carbon produced after each cycle was characterized by nitrogen adsorption isotherms using High-Pressure Volumetric Analyzer. Micropore volume fraction and pore size distribution showed insignificant changes whereas the BET surface area slightly decreased approximately by 5% from 2610 m^2/g up to the 10th cycle. To assess the applicability of the activated carbons developed through the recycling KOH process, hydrogen storage was analyzed in the resulting activated carbon of each cycle and substantial hydrogen uptake was found for 10 cycles. Through this study, the applicability of KOH recycling by using the novel technique of maintaining a constant pH of the activating solution is highlighted where the surface morphology insignificantly varied between the cycles and hence the applicability in hydrogen storage capacity of the material demonstrated an inconsequential effect.

632128 Implementing Autothermal Reactions in a Novel Fluidized Bed Model for Fast Pyrolysis Applications

Abstract:

Recent developments have identified biomass as a viable, renewable alternative to petroleum-derived fuels and feedstocks through fast pyrolysis reactions. In particular, the implementation of autothermal operation has overcome the major limitation of supplying externally generated heat to the reactor itself. By coupling the pyrolysis reactions with combustion of char and other less valuable products, the biomass throughput a fluidized bed pyrolysis reactor has been shown to increase five times as compared to normal pyrolysis.

This work builds on prior work, developing a model for autothermal pyrolysis in a novel fluidized bed reactor. The rapid evolution of pyrolysis vapors prevents competition between pyrolysis and oxidation reactions, allowing straightforward introduction of oxidation reactions into the existing model. Preliminary results are in good agreement with the available experimental data.

628188 Double the Pleasures: Perennial Grasses As Soil Phytoremediation Agents and Renewable Carbon Resources for Fuels and Chemicals

Abstract:

Biomass is the only renewable source of carbon that, when utilized efficiently and optimally, can significantly help reducing our reliance on fossil-based carbon resources. Lignocellulosic biomass is abundant, inexpensive and does not compete with the production of food crops. Nevertheless, the complex nature of biomass and many ill-defined issues related to biomass utilization pose a substantial challenge to large-scale biomass utilization. Growing biomass for energy crops means that more land needs to be utilized. It is desired that biomass for energy crops would be grown in land areas that are considered marginal and not currently being utilized for anything productive.

Certain types of fast-growing perennial grasses, such as switchgrass, have great potential to be used as renewable feedstock to produce bioenergy due to their relatively high yields of biomass per acreage of land. These grasses can grow well in land areas which are otherwise not suitable for conventional farming. They also can resist a wide range climate temperature, thus can grow in many parts the world including in cold regions. The fast-growing characteristic, along with the ability to grow in marginal land, make these perennial grasses highly potential sources for renewable carbon.

Another unique characteristic of switchgrass is that the plant spreads by short rhizomes and its roots can go deep more than three meters below the surface. The rhizomes and deep roots make switchgrass a potential plant to be grown in soils that are contaminated with heavy metals. As a phytoremediation agent, switchgrass can extract these heavy metals from the contaminated soil, which in turn can improve the quality of the soil. This method is potentially much less expensive than the conventional soil remediation method, such as ex-situ excavation method, which is a highly energy intensive process.

In this study, the potentials of switchgrass as soil remediation agent and as feedstock for a biorefinery system were evaluated. A pot study of growing switchgrass in soils that were doped with lead at various concentration demonstrated that the grass could grow in the contaminated soils and extract the heavy metal. Germination and growth rates of the grass were not affected when the lead concentration in the soil was between 0 – 200 ppm. Based on data, it was estimated that it will only require 3.5 years to reduce the lead concentration in the soil from 500 mg Pb/Kg of soil to 10 mg Pb/Kg of soil. The lead-contaminated switchgrass plants were harvested, dried, and comminuted into millimeter-size particles to be used as a feedstock for a biorefinery process system. Four process routes consisted of enzymatic hydrolysis, acid hydrolysis, fast pyrolysis and their combinations were evaluated to produce sugars, bio-oil, and biochar.

The study showed that the presence of lead in the grassy feedstock did not seem to affect the yields sugars and bio-oil from the hydrolysis and the fast pyrolysis steps, respectively. An LCA study on growing switchgrass for soil phytoremediation and using the switchgrass as feedstock for the biorefinery system was performed. Results from the study showed that the combination of using switchgrass as soil phytoremediation agent and feedstock for producing energy in a biorefinery decreases GHG emissions and lowers the use of energy produced from fossil fuels when compared with a combination of a refinery system and ex-situ excavation for soil remediation.

624983 Conversion of Biowaste to Value-Added Biochemicals and Functional Biomaterials for High-Efficient Circular Economy

Abstract:

Our society is facing long-term environmental challenges such as extreme climate change, resource and energy depletion, air pollution, food scarcity, water and soil contaminations. To tackle these environmental and sustainable concerns and challenges, our study focuses on the use of green chemistry and engineering technique to produce value-added bioproducts including platform chemicals, functional biomaterials, and biofuels from abundant renewable resources, especially low-cost waste. In this talk, the major research areas in four aspects will be introduced at first: 1) Fabrication of functional nanomaterials for mitigating water, soil, food contaminants; 2) Synthesis of chemical building blocks & monomer precursors for sustainable polymer systems; 3) Catalytic conversion of lignin to high value aromatic platform chemicals and hydrocarbon fuel precursors; and 4) Sustainable process modeling and optimization. This talk will present the overall studies on the development of a wide range of value added bioproducts in the applications of food safety, medical, environmental applications. The key problems during the self-assembling, molecular synthesis, and catalytical valorization will be also elucidated by several recent works.

625084 Microencapsulation of Flame Retardants Using Bio-Based Materials

Abstract:

The development of non-toxic and durable flame retardant fabrics using surface treatment methods is of interest for numerous applications. However, most approaches currently reported for surface treatments render the fabric stiff and non-breathable. Microencapsulation is a strategy that can be used to incorporate additives onto textiles. We present an approach to encapsulate flame retardants using safe bio-based materials and coat them onto a Nylon-Cotton (NyCo) fabric surface. A phosphorous-based flame retardant, melamine polyphosphate, is encapsulated using bio-based materials like gelatin and tannic acid, a plant-based polyphenol. The process parameters such as concentration of ingredients, reaction time, and stirring rate are varied to optimize encapsulation efficiency and flame retardant properties. The microcapsules are characterized using Scanning Electron Microscopy (SEM), Energy Dispersive Spectroscopy (EDX), Particle Size Analyzer, and FTIR-ATR. The thermal stability and heat release characteristics of these capsules are studied using Thermogravimetric Analysis (TGA) and Pyrolysis Combustion Flow Calorimetry (PCFC), respectively. The flammability of microcapsule-coated fabric swatches is studied using a Vertical Flame Tester (VFT). Flammability data shows that the microcapsule coating imparts flame retardant characteristics to the fabric. Use of safe flame retardants and encapsulating materials opens new possibilities for safer and durable flame retardant solutions for fabrics using a microencapsulation approach.

625381 Impacts of High Shear-Rate Processing on Cellulose Nanofibrils

Abstract:

Cellulose nanomaterials provide a promising avenue for combining composite materials with natural polymers or bio-based plastics, however current studies on rheological characterization at high shear rates is limited. In this study, a capillary rheometer was employed to investigate the rheological characteristics of aqueous nanofibrillated cellulose suspensions (NFCs) at room temperature and shear rates above 1000 s^{-1}. The shear rate-dependent viscosity of NFCs was characterized and monitored for viscosity-related indicators of fibril decomposition or entanglement due to the stresses induced by the high shear-rate flow. Traditional capillary rheology corrections for entrance pressure loss and Non-Newtonian flow profiles were attempted to compare apparent and true rheological behaviors. Large entanglements of fibrils made the characterization of the suspensions difficult due to their ability to clog the capillary, particularly with fine capillary dimensions. However, appropriate well-characterized preprocessing steps enable significantly improved measurement resolution and could be implemented on a large scale to enhance rheology and modeling analyses for fibril distribution in composite materials.

624761 Chitosan-Nanoparticle Enhanced Antibiotic and β-Lactamase Inhibitor to Treat Multi-Drug Resistant Pathogens

Abstract:

Antibiotic resistant bacteria (ARBs) are responsible for at least 2,868,700 infections and 35,900 deaths annually in the United States. Among ARBs, one type that causes serious threats to public health is extended-spectrum b-lactamase (ESBL)-producing *Enterobacteriaceae*, which mediates the resistance to most b-lactamase antibiotics (e.g penicillins) and undermines the efficacy of current antibiotic therapies. Thus, it is critical to make antibiotics effective again through low-cost and high-effective nano-antibiotics or alternative antimicrobials. In this study, we develop an innovative antimicrobial therapy against multi-drug resistant bacteria by engineered chitosan nanoparticle coated β-lactam antibiotic and β-lactamase inhibitors (CNAI) using a water-oil-water (w/o/w) emulsion method. The w/o/w ensures the high-effective encapsulation of chitosan nanoparticles on the antibiotic and inhibitors and a slow and controllable drug release via a core-shell nanostructure. The chitosan nanoparticles with antibiotics/inhibitors (CNAI) were evaluated for their physicochemical properties, antimicrobial activity, and the drug loading behaviors. The results show that all combinations of CNAI have a particle size of around 200 nm, a Minimum Inhibitory Concentration (MIC) of 1 ug/ml and are effective at inhibiting multi-drug resistant bacteria. The chitosan nanoparticles engaged in the system are expected to enhance antibacterial properties and provide controllable release function to increase the entire antibiotic efficiency.

627133 Tackling Water Pollution Via Sustainable Sorbents: Pb(II) Adsorption By Bio-Based Carbons Via First Principles Computational Models

Abstract:

Accessing clean water—one of the critical components of life and irreplaceable in the agriculture and energy sectors—is currently a global crisis that is growing in severity as the effects of climate change worsen. Bio-based carbons produced from the upgrading of lignocellulosic biomass waste present an opportunity to develop highly efficient and inexpensive sorbents for extraction of heavy metals from watercourses. Adsorption studies have demonstrated that these sustainable materials are three times as effective at removing copper, lead, zinc, cadmium, and cobalt than a more traditional sorbent, i.e. activated carbon. However, the complex, amorphous nature of bio-based carbons makes it nearly impossible using experiments alone to determine the fundamental, atomic-scale chemical structures, phenomena, and mechanisms that lead to efficient binding of heavy metals. Thus, there is a significant challenge here-in that the design and optimization of highly efficient bio-based carbons is limited by a lack of fundamental chemical insights into the chemical structures present on bio-based carbons and their subsequent ability to interact with and bind heavy metals. We address this challenge here by combining atomic-scale computational models (i.e. density functional theory) with fundamental, statistical mechanics to (1) determine the dominant functionalities present on bio-based carbon sorbents and (2) probe the Pb(II) adsorption mechanism onto these dominant functionalities.

First, to determine the dominant functionalities on bio-based carbons, a database of 225 distinct carbon, hydrogen, and oxygen containing defects, with the latter including hydroxyl, carboxylic, and ether functionalities, was created in a model graphene surface. The formation free energies for each defect structure were calculated at a range of temperatures (i.e. 200-2000 K) and gas phase oxygen and hydrogen chemical potentials, resulting in a series of phase diagrams that show the most thermodynamically stable—i.e. dominant—defects across a range of environmental conditions. For example, at 1000 K (Figure 1A), only four of the 225 tested defects are thermodynamically accessible, with the dominant defect structures being the Stone-Wales defect under highly reducing environments and a double carbon vacancy with four ether groups under highly oxidizing environments. Thus, our rigorous scan of the bio-based carbon defect space enables a physics-guided reduction of possible chemical structures to a database of those structures most likely to be present on bio-based carbons.

Second, to probe the Pb(II) adsorption mechanism onto the dominant functionalities, a Pb(II) complex was adsorbed to each dominant defect. On the defects containing only carbon, the Pb(II) complex adsorbs weakly (~-15 kJ mol^{-1}) and predominantly via van der Waals forces. The addition of oxygen to the adsorption site significantly increases the Pb(II) adsorption strength (~-86 kJ mol^{-1}). As shown in Figure 1B, the presence of oxygen defects in the model bio-based carbon creates a localized, negatively charged site unto which the positively charged Pb(II) complex bind. Thus, the structure and number of defects, whose formation can be directed using the atomic-scale insight from our phase diagrams, controls the sorption capacity of the bio-based carbons. Overall, the linkages elucidated here between bio-based carbon sorbent structure and Pb(II) adsorption capacity establish the necessary fundamental chemical design rules to facilitate the rational design of highly efficient bio-based carbon sorbents for water purification.

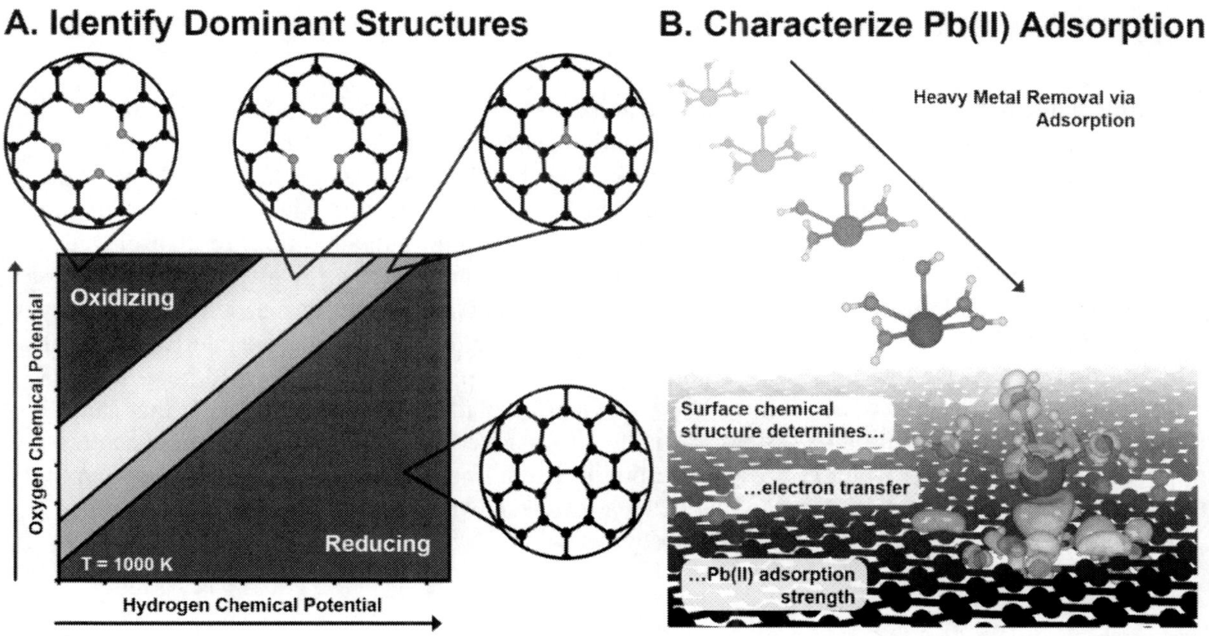

Figure 1. *(A) A rigorous scan of the thermodynamic stability of carbon, hydrogen, and oxygen defect space within a graphene model of bio-based carbon enables the identification of the dominant defect structures present under a wide range of temperatures and gas phase environments. (B) Each structure is then evaluated for its ability to bind heavy metal complexes at the atomic-scale, allowing for construction of predictive chemical structure-function relationships. In the ball and stick models, the black, red, green, and dark gray spheres represent carbon, oxygen, hydrogen, and Pb(II), respectively. Additionally, the yellow (blue) volumes in (B) represent the electron density gained (lost) upon adsorption of the Pb(II) complex onto the bio-based carbon model.*

629058 Dewatering of Cellulose Nanofibers Using Ultrasound

Abstract:

A high-priority goal for advancing the manufacture and broad applications of cellulose nanofibers (CNF) is to address the high-energy requirements associated with dewatering and drying of these materials. Additionally, the current dewatering and drying methods lead to the unwanted issues of agglomeration (loss of nanoscale dimensions), negatively impacting their physical properties. Thus, there is a need for developing an energy efficient dewatering process for CNF with minimum alterations to their physical properties. In this talk, we will describe a new low-cost, low-heat and low-energy method of dewatering CNF based on ultrasound technology. Ultrasonic waves generated using ceramic piezoelectric transducers are applied to the CNF slurry, leading to the formation of instabilities at the liquid-air interface, cavitation, and the subsequent atomization of the trapped liquid from CNF slurries in form of a cold mist. To achieve high throughput, we will discuss a continuous flow dewatering platform as shown in Fig 1, that can facilitate scaleup and optimization of system parameters for desired final dewatered slurry concentration. Our platform achieves ~70% water removal corresponding to a final solid weight of ~10% from an initial value of ~3% of CNF slurry. We also find that our platform achieves significant reduction in energy consumption as compared to other drying methods.

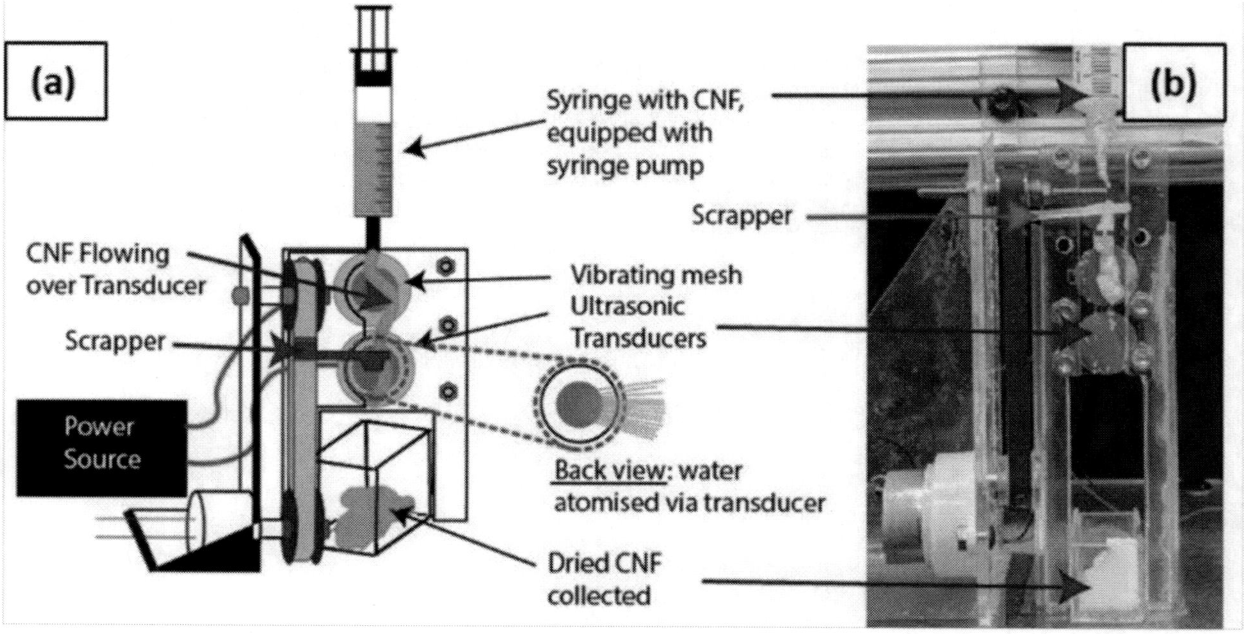

Fig:1 a) Schematic corresponding to the experimental setup for the continuous mode ultrasonic dewatering of **(b)** Image of the experimental setup .

631089 Activated Carbon from High-Purity Lignin with Controlled Molecular Weight

Abstract:

Over the past decade, the production of biofuels from lignocellulosic biomass has steadily increased to replace fuels from petroleum sources. To make these biofuels cost competitive, it is essential to create products from the "ligno-" components (up to 30% by mass) of the biomass, which requires separating and purifying the byproduct. Activated carbon (AC) is a viable product from the biorefineries since it has a market capacity large enough to consume the necessary amount of lignin produced, and it does not have a stringent requirement for its precursor. However, the relationship between lignin properties and its derived AC properties are not fully understood, especially the effect of lignin molecular weight (MW) and ash content on the key AC properties.

In this work, we use various fractions of ultraclean lignin (from a patented ALPHA process) with different MWs to study how lignin MW and ash content affects the key properties of AC using controlled activation conditions. The pore structure of the ACs were characterized by N_2 adsorption at 77K using BET analysis and the aqueous adsorption capacities were characterized by using iodine number and methylene blue value. The experiment results show the surface area and the adsorption capacities of the ACs increased when using lignin precursors with increased MW, and the purification of the lignin can significant enhance the AC properties. The surface area of the AC from the highest MW lignin can reach to ~1900 m^2/g and still has potential to be improved. The high purity and the flexible MW could facilitate the co-produced lignin to produce AC with desired pore structures which can be applied in broad applications. Furthermore, the proposed method in this work for synthesizing AC is not complicated and therefore, is suitable for bulk production.

623222 Production of High-Quality Carbon Fiber from Lignin Precursor

Abstract:

Lignin is considered as a promising precursor material for producing low-cost green carbon fiber due to its abundant availability at low costs and high aromatic content. Unfortunately, lignin-based carbon fibers are difficult to commercialize due to their poor mechanical properties. Lignin-based carbon fiber is usually made using a fabrication process previously developed to convert polyacrylonitrile into carbon fibers, despite that the precursor materials have very different chemical structures. The questions regarding lignin-based carbon fiber, such as how the physical and chemical properties of lignin change during the fiber oxidation and subsequent carbonization, and how these property changes affect the mechanical properties of the final products, remain unclear to date. Due to the lack of the fundamental insights into lignin transformation, previous approaches to improve the quality of the carbon fiber has been largely unsuccessful. In this study, carbon fibers were produced from softwood-derived organosolv lignin using a melt-spinning method. The fibers at different stages of the fabrication were carefully investigated using comprehensive analytic techniques to elucidate the relationships between the chemical structure and material properties of the fiber relevant to the fiber fabrication conditions. The results showed that applying tension can significantly modify the chemical structure of the lignin fiber during oxidation and carbonization, thus leading to carbon fibers with unique carbon structures and strong tensile properties. In our study, the average tensile strength and tensile modulus of the carbon fiber reached 1.96 GPa and 158 GPa with a carbonization temperature of 700 °C, and 2.12 GPa and 164 GPa with a temperature of 800 °C, respectively. Our research finding suggested that a good understanding about chemical structure and material property correlation is a key to develop high-quality carbon fiber using lignin.

626307 Exploiting the Liquid-Liquid Phase Behavior of Hybrid Poplar Lignin in Ethanol-Water Solutions to Produce Precursors for Value-Added Applications

Abstract:

Lignin is an abundant biopolymer with unique functionality that is a commonly produced byproduct from cellulose enrichment processes such as papermills. This aromatic polymer is synthesized by plants to add structural support and rigidity to woody biomass. Conventionally, papermills burn lignin for its heating value in order to cover the energy demands of the mill while reclaiming catalytic salts for the pulping process. While these mills are well established, bioethanol refineries are generally recognized as unprofitable, even if they could potentially produce fuels with a lower environmental impact. If the coproduct lignin could be recovered, purified, and sold at a profit, the value proposition of bioethanol refineries would be fundamentally improved. On top of supporting these refineries, finding economically feasible and renewable means of purifying/upgrading lignin opens up this currently under-utilized resource as a new feedstock for material and chemical applications. Previous work with Kraft lignins has shown that ethanol-water solvents (with specific composition, temperature, and solvent-to-lignin ratio) produce liquid-liquid phase equilibria. Titled as the ALPHA (Aqueous Lignin Purification with Hot Agents) Process, this technology has been applied to lignin sources isolated from Hybrid Poplar (HP) trees via alkaline pulping. Ethanol-water solutions have proven to be effective at simultaneously fractionating HP lignin by molecular weight and removing impurities that are harmful to downstream, value-added applications. In this work we examine the impacts of fractionation and purification on the ability for HP lignin to serve as a precursor for applications such as carbon fibers, polyurethane foams, and activated carbon.

629900 Synthesis of Charged Lignin Nanoparticles and Its Applications As Adsorbent

Abstract:

Per- and poly-fluoroalkyl substances (PFAS), which are also called forever chemicals, are widespread in surface water, groundwater and drinking water in the United States. They are recalcitrant to present treatment techniques, accumulate in the human body over time and cause serious health problems. Challenges for the removal of PFAS from the water are low adsorption capacity, slow adsorption, and high cost of the adsorbent. Lignin, the most abundant low-cost aromatic polymer, is an underutilized polymer due to its heterogeneous structure. Lignin nanoparticles (LNPs) offer many opportunities for value-added applications due to their controllable size, spherical shape, large surface area, high surface charge, long-term dispersibility in water, and compatibility with other polymers. Since the majority of PFAS contain negative charges on their surface, electrostatic interaction between PFAS and adsorbent can be increased by synthesizing cationic adsorbents. In this study, cationized lignin was prepared and applied for PFAS capturing. Adsorption of PFAS was compared for technical lignin, cationic lignin, and cationic LNPs. Regeneration of lignin particles was performed by increasing the pH. The structural, morphological and adsorption properties were analysed and discussed using GPC, HPLC, FTIR, ^{1}H NMR, 2D HSQC NMR, ^{31}P NMR, zetasizer, SEM, and others.

625070 Techno-Economic Analysis and Life Cycle Assessment of Lignin Fractionation and Valorization Via the ALPHA Process: Upgrading to Value-Added Products

Abstract:

This study investigated the environmental and economic sustainability of the Aqueous Lignin Purification with Hot Agents (ALPHA) process. The ALPHA process has been recently investigated for the fractionation and purification of raw, bulk lignins recovered from cellulosic ethanol biorefineries or Kraft pulp mills. The ALPHA process exploits the novel liquid-liquid equilibrium that exists between lignin and hot, one-phase solutions of aqueous renewable solvents to simultaneously purify and fractionate raw bulk lignin. In particular, both the purity and molecular weight (MW) of a given lignin fraction can be controlled. Here, ALPHA is proposed for the isolation of lignin from a corn stover-to-ethanol plant into purified low, medium, and high MW fractions for producing polyurethane foam, activated carbon, and carbon fiber, respectively. Two aqueous renewable solvents were analyzed: acetic acid and ethanol.

Our analysis shows positive economic and environmental results for producing lignins of controlled molecular weight and purity via ALPHA. Solvent choice was found to have a significant impact on production of fractionated lignin using ALPHA, with a minimum selling price of $838/tonne for using acetic acid vs. $463/tonne with ethanol. Converting lignins processed with ethanol–water solutions into high-value products such as activated carbon, carbon fibers, and polyurethane foam yields $151 million/year in profit, which over 30 years results in a total net present value (NPV) of $533 million using a discounted cash flow analysis. The NPV is most sensitive to lignin yield, solvent choice, and the lignin substitution percentage for polyol in PU foam. A life cycle assessment was conducted to determine the "gate-to-gate" greenhouse gas emissions and energy consumption of the lignin-based products compared to fossil-based equivalents using SimaPro. A value allocation scenario was conducted and determined that products generated using the ALPHA process with ethanol have similar or lower greenhouse gas emissions than the same products from fossil feedstocks.

628071 Characterization of Novel Soybean Hull-Based Binders for Aqua-Feed Pellets

Abstract:

Soybean is one of the most cultivated crops in the world generating about 18–20 million tons of soybean hull (SBH) per year in the world (with ~50% of it for US only). Current applications of SBH are mainly animal feed and dietary fiber, with the rest being landfilled or incinerated. With the goal of producing value-added products from SBH and reducing the environmental problems caused by landfilling or incineration of this under-utilized biomass, the Alabama Center for Paper and Bioresource Engineering (AC-PABE) has recently developed a novel compound feed binder using 100% soybean hull (SBH) in collaboration with USDA-ARS [1]. The novel binder has exhibited superior binding performance in laboratory scale evaluations as compared to the commercial feed binders: carboxymethyl cellulose, corn starch, and wheat gluten [1]. In addition, the novel binder's production requires only traditional, simple to apply, and scalable acid treatment with minimum effluent.

The objective of this study is to characterize the chemical and physical properties of the novel binders and understand the interactions among different components such as protein, pectin, hemicellulose and cellulose and their synergistic effects under different processing conditions, with a goal of identifying the optimal processing conditions.

Evaluation of the mass balance of the filtrate and retained solids from the acid treatment showed that a significant amount of non-fibrous components, i.e., pectin and protein, are solubilized during the treatments (Fig 1). Chemical composition and functional group analysis confirmed the dissolution of pectin and protein, contributing to the binding properties of the SBH binders (SBH-1, 2, 3) as opposed to the untreated SBH (Table 1). The carboxylic acid content for SBH-1 is twice the value for the untreated SBH. This value is even higher for SBH-2 and SBH-3, due to the separation from fibers and having more dissolved pectin and protein in them (about 18% compared to 10% in SBH-1). We are evaluating the structural changes of pectin, protein, SBH fibers, and their synergistic effects under different processing conditions.

[1] B. Aksoy, Z. Jiang, M. Aksoy, B. Beck, Animal feed binders derived from pectin-and protein-containing feedstock, US Patent App. 16/745,562 (2020).

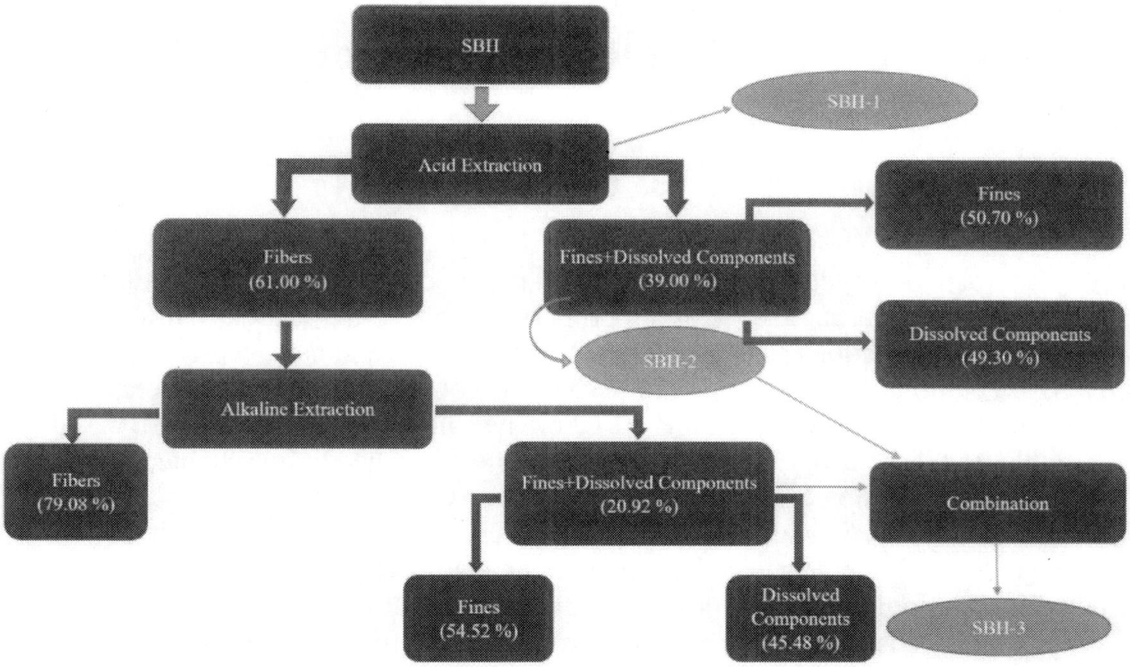

Fig 1. Patented production of SBH binders and the corresponding mass balance [1]

Table 1. Chemical composition and functional group analysis of untreated SBH and SBH binders

Sample	Carboxylic Acid Content (mmol/100 gr)	Protein Content (%)	Pectin Content (%)
Untreated SBH	43.75±0.32	10.34±0.01	11.68±0.07
SBH-1	86.23±2.86	10.09±0.03	10.58±0.02
SBH-2	100.98±2.72	18.04±0.06	18.88±0.14
SBH-3	89.76±2.24	17.59±0.11	18.34±0.12

633847 Biodegradable Plastic Blends from Polyhydroxyalkanoate and Cellulose Ester for Sustainable Packaging

Abstract:

Poly(3-hydroxybutyrate-co-3-hydroxyvalerate) (PHBV) is a biodegradable, biocompatible, bio-based bacterial polyester, making it attractive for use in primary packaging applications. However, for flexible packaging, the applications are limited due to its brittleness and narrow processing temperature range. To mitigate the brittleness of PHBV, the compostable cellulose acetate (CA) may improve the toughness of PHBV. Furthermore, based on the solubility parameters, PHBV and CA are potentially miscible. In this research, PHBV/CA blends were fabricated to investigate the affect of toughness on CA and PHBV, however CA requires plasticization in order to be melt compounded. As such, CA powder was plasticized with a non-toxic triethyl citrate (TEC). Subsequently PHBV pellets were melt compounded with the plasticized CA powder in a twin-screw extruder followed by injection moulding. Our findings suggest possible degradation of TEC and PHBV, due to the high processing temperature, as determined by a thermogravimetric analysis (TGA). Scanning electron microscopy (SEM) micrograph and density measurements revealed increased porosity of PHBV/CA blends. Blends with high plasticized CA have improved impact of 110% relative to virgin PHBV. Different processing temperatures were also assessed to investigate the different properties of the material and the heat deflection temperature (HDT) of the blend which declined by approximately 10 °C when processed at a higher temperature.

Acknowledgment

This research is financially supported by the Natural Sciences and Engineering Research Council (NSERC), Canada Discovery Grants (Project # 400320); the Ontario Ministry of Agriculture, Food and Rural Affairs (OMAFRA) – University of Guelph, the Bioeconomy Industrial Uses Research Program Theme (Project # 030255); and the Ontario Ministry of Economic Development, Job Creation and Trade ORF-RE09-078 (Project #053970).

632963 Synthesis of Biobased Phenol-Formaldehyde Wood Adhesives from Biorefinery Derived Lignocellulosic Biomass

Abstract:

The high utilization of adhesives in the wood industry and increasing use of petroleum-derived chemicals for their synthesis has enhanced the opportunities for lignocellulosic biorefineries to reduce environmental and energy crises. In this study, lignin received from kraft biorefinery, bio-oils obtained from solvent liquefaction, and fast pyrolysis, laboratory-scale thermochemical conversion processes have been thoroughly characterized by FTIR, 2D HSQC-NMR, and GCMS technique before utilizing them as a monomer in phenol-formaldehyde resin synthesis. To further understand the structure of PF (phenol-formaldehyde) resins, FTIR and ^{1}H-NMR analysis were performed. Additionally, the reaction of PF resin with hexamethylenetetramine curing agent favors the crosslinked structure, and this curing phenomenon was studied using differential scanning calorimetry (DSC), where every resin and curing blend shows a deep exotherm peak during the process. The adhesive performance of resins on the wood substrate was evaluated by shear bonding strength under both wet and dry conditions along with physical properties such as water absorption and swelling behavior. The result from the study shows that incorporating kraft lignin was able to improve the bonding strength of wood specimens.

638816 Waste wheat starch-based home compostable plastics for packaging applications

Abstract:

The accumulation of waste plastics and pollution caused by improper disposal are among the most pressing issues threatening the health of ecosystems around the world. While starch is abundant, inexpensive, renewable, and biodegradable, it cannot be used directly as a plastic. This study takes waste wheat flour from the food industry and combines it with a biodegradable plastic to produce bioplastics and composites that are home compostable at the end of their life. After plasticizing the waste starch with glycerol (recovered as a by-product of the bio-diesel industry) to produce plasticized wheat starch (PWS), thermoplastic starches were successfully prepared through with poly(butylene adipate terephthalate) (PBAT) in a 30/70 ratio. Extrusion and injection molding techniques were used to produce test specimens in accordance with ASTM standards. To enhance the mechanical and thermal properties, calcium carbonate (CaCO3) and talc were added as fillers to the thermoplastic starch. The melt flow properties, thermal behaviour, surface functionality, morphology, and dimensional stability at high temperatures were analyzed along with the mechanical properties. PBAT/PWS (70/30) specimens displayed the highest elongation at break values of 208.28%, making the formulation suitable for packaging applications. The addition of CaCO3 resulted in an improved melt flow rate of 27 g/10min while the talc powder had a value of 14 g/10 min. The talc composite showed the highest tensile modulus of ~475 MPa, 134% higher than that of PBAT/PWS (70/30). The waste-starch-based composites show promise as sustainable alternatives to single-use fossil-fuel-based plastics with reduced environmental impact.

Keywords: Plasticized starch, starch bioplastics, starch-based composites.

Acknowledgements: (i) The Ontario Ministry of Agriculture, Food and Rural Affairs (OMAFRA) – University of Guelph, the Bioeconomy Industrial Uses Research Program Theme (Project Nos. 030486 and 030578); (ii) OMAFRA-University of Guelph Gryphon's Leading to the Accelerated Adoption of Innovative Research (LAAIR) Program (Project No. 030416); (iii) the Ontario Research Fund, Research Excellence Program; Round-9 (ORF-RE09) from the Ontario Ministry of Economic Development, Job Creation and Trade, Canada (Project No. 053970); and (iv) the Natural Sciences and Engineering Research Council of Canada (NSERC), Canada Research Chair (CRC) program Project No. 460788.

629309 A Process to Utilize Rice Straw for High-Value Acoustics Applications

Abstract:

Agro residues are produced in considerable amounts worldwide which are dumped to landfills or burnt for heating with very low energy recovery. On the contrary, these agro residues have very high potential to be used as feedstock in some of the high value applications which includes the fabrication of acoustic materials in the form of microfibrils composed holocellulose. However, the complex arrangement of lignin, cellulose, and hemicellouse in lignocellulosic biomass should be broken before their utilization in acoustic materials. Therefore, present study also focusses on the delignification of rice straw to produce microfibrils for acoustic applications. Production of microfibrils for acoustic applications provide an alternate high value, innovative and cost effective technique for rice straw utilization compared to conventional agro residue conversion processes. The efficacy of the proposed technique could be quantified on the basis of cost competitiveness of the final product compared to commercially available alternatives, and reduction in propensity of waste generation during conversion of rice straw into acoustical materials.

In this study, NADES were derived from biorenewable resources that needs low-cost recycling, shows high reusability, non-toxicity, and mild acidic properties, have been synthesized and used for the delignification of rice straw. Firstly, NADES was prepared by incubating a mixture of lactic acid (LA), choline chloride (CC) and water in 9:1:0.11 molar ratio at 60-80°C for 30 minutes in a reactor. After that, synthesized NADES was characterized using melting point temperature

(Tm), and glass transition temperature (Tg) analysis to validate the formation of desired NADES. Further, a clear solution from incubating LA-CC-water at aforementioned conditions also indicated NADES formation. On the other hand, rice straw was grinded to powder form and sieved to obtain the particles of less than 0.5 mm size. Post this, delignification of powdered and sieved rice straw was carried out using natural deep eutectic solvents (NADES). In general, 5-10 wt% of powdered and sieved rice straw was pre-treated with NADES (LA, CC and water) at 120°C for 60-200 minutes to produce cellulosic microfibrils. Notably, NADES due to their mild acidity need to be removed from the microfibrils before fabrication of the final product, therefore, microfibrils were washed by distilled water until the neutral pH (6.8-7.2) was obtained. Eventually, obtained microfibrils were used for acoustic materials fabrication via different techniques and the efficacy of the sheets has been examined on impedance tube set-up.

The results suggest that the sheet made from rice straw microfibrils could be utilized in a range of acoustical applications and have effectiveness comparable to commercially available sheets made of synthetic materials. It has been found that the acoustical performance of sheets made from rice straw improves significantly after conversion into microfibrils due to reduced fiber diameter, and increased tortuosity. Further, reusability of NADES for multiple cycles of operation without loss of pretreatment efficiency leads to significant reduction in pretreatment cost and ultimately enhances the profitability of the entire process.

634535 Properties and Applications of Biofoam Composite Materials

Abstract:

Agricultural materials are of interest as a replacement for plastics since they are typically inexpensive, widely available, renewable, and compostable. Biofoam is of particular interest as a potential replacement for plastic foam products such as those made of polystyrene. Biofoam with uniquely different properties can be made using a variety of processes including solvent exchange, extrusion, thermoforming, and mixing techniques among others. Biofoam composite materials that will be discussed include microcellular starch foam, thermoformed starch/fiber composites, and fiber-based semi-rigid foam panels with insulative and cushioning properties.

634155 Sustainable Plastics at NRC: Towards Zero Plastic Waste and Circular Economy

Abstract:

How do we shape future industrial materials towards zero plastic waste and circular economy?

The National Research Council of Canada (NRC) is working intensively on the development of environmentally friendly plastic products for industrial applications such as packaging, transportation, and medical devices. This presentation will cover some of the research areas investigated by the NRC, including the valorization of forest and agricultural residues into polymer products, the development of compostable products, and the conversion of plastic waste into useful materials or molecules.

634055 On Utilization of Agrowastes for Biocomposites and Biochar

Abstract:

The annual demand for wood and wood composites is increasing steadily. It is expected that the worldwide demand for solid wood and products would continue to increase by 60% by the year 2030 and the demand is expected to continue through the year 2060. The price of wood is continuously increasing with the demand. There is a great opportunity and the need to utilize agro-based and forest waste-based particle boards, or Oriented Strand Boards (OSB). Defatted soymeal, and distillers dried grains with solubles (DDGS) can provide cost-competitive and environmentally friendly biobased adhesives for making the wood composites. Properly designed biobased composites can provide desired physical and other mechanical properties. Plant fibers from corn and sorghum stalks can also be used in OSBs. These composites can be designed to improve thermal insulation and weathering properties. Wastes from the hardwood lumber industry and the biobased adhesives can be incorporated into the particle boards, without the requirement of major process changes. Similarly, agro-wastes such as soybean hulls, avocado pits, and waste railroad ties can provide raw materials for producing large quantities of biochar. Biochar can replace petroleum-based carbon in water treatment, agricultural soil amendments, and process separations. The biochar can help with carbon sequestration, as it reduces the net release of carbon dioxide and also reduces the release of potent greenhouse gas, methane. The presentation includes select examples of biobased composites and approaches for biochar preparation and applications.

636348 Circular tools, materials and business models for a circular economy

Abstract:

As we strive for a circular economy, companies are looking for ways to close the loop on their products. At Stora Enso Circular Solutions, we are developing tools and business models to utilize more recycled raw materials and ensure that our products are re-used and recycled as much as possible. For utilizing recycled raw materials, novel sourcing methods are needed. We are developing a digital platform concept together with partners that would enable better visibility and connectivity to those seeking to sell, buy or help upcycle recyclable raw materials. Of course, sourcing of recycled raw materials is not the end game, and we are also working on methods of using bio-based ingredients to upcycle and improve CO2 footprint of recycled plastics. Ultimately, we need to make money from these materials as well. But ensuring the maximum value of our materials means closing the loop and taking responsibility beyond the material production. We are developing new business models for owning and leasing products made from our biocomposite materials including a closed loop for 3D printed furniture. With these developments and more to come, we look forward to contributing perpetually towards a circular economy.

633850 Waste Valorization and Sustainable Biocomposites Towards a Circular Economy: Current Status and Future Opportunities

Abstract:

Innovative and sustainable solutions are required to combat the world's growing waste and pollution problem which is contributing to GHG emissions. Currently, the linear economic model of "make-use-dispose" is rapidly increasing the amount of generated waste and our ability to process it in a way that doesn't harm the environment. A closed loop approach of managing resources is essential. Developing new products is an integral part of this approach, also called a circular economy concept. New products should be designed to minimize waste or utilize waste as a resource. This presentation will focus on valorizing waste from various industries such as plastic, agriculture, food, biorefining, manufacturing, automotive, and packaging to develop sustainable biobased plastics composites. The incorporation of waste from these industries provide potential value-added uses in materials, if appropriately integrated in the design and engineering. Plastic makes up the majority of the municipal solid waste, where only about 9% is recycled, while the remainder is incinerated or ends up in the landfill. Similarly, the United Nations Food and Agriculture Organization (FAO) approximates that 1/3 of the food produced - 1.3 billion tonnes - goes to waste which is estimated to cost $990 billion US per year. Highlights of this presentation will include the use of recycled plastics and industrial wastes (including food processing waste) as renewable feedstocks for innovative industrial applications. Wider applications of these materials will help lessen the waste entering the landfill, therefore reducing pollution and GHG emissions.

Acknowledgements: (i) The Ontario Ministry of Agriculture, Food and Rural Affairs (OMAFRA) – University of Guelph, the Bioeconomy Industrial Uses Research Program Theme (Project Nos. 030486 and 030578); (ii) OMAFRA-University of Guelph Gryphon's Leading to the Accelerated Adoption of Innovative Research (LAAIR) Program (Project No. 030416); (iii) the Ontario Research Fund, Research Excellence Program; Round-9 (ORF-RE09) from the Ontario Ministry of Economic Development, Job Creation and Trade, Canada (Project No. 053970); and (iv) the Natural Sciences and Engineering Research Council of Canada (NSERC), Canada Research Chair (CRC) program Project No. 460788.

627787 Natural Fiber Composites Using Seed Hulls from Grain Processing

Abstract:

Utilization of agriculture process residues (e.g., soy hulls from soybean processing; corn fiber from distilleries and ethanol plants; rice hulls from rice milling) for the production of sustainable and value-added co-products are critical for improving environment and bioeconomy. The hulls from grain processing are globally available in large volumes at a consistent quality. Their shipping, storage, and other logistics costs are more effectively manageable compared to other traditional biomass feedstocks.

In our studies funded by United Soybean Board (USB), our team at University of Louisville utilized soy hulls to separate and modify hull fibers for natural fiber composites; while simultaneously producing xylose (natural sugar) as a value added product. This approach not only reduced the cost of fibers; but also produced fibers that are more compatible with polymers which helped reduce air traps occurring at fiber-polymer interface leading to improved material properties of natural fiber composites. Using our integrated approach, with xylose (market value \$3/kg) as a co-product, we expect the production cost of natural fiber composites to be competitive. Wide use of low cost, light weight, and high performing natural fiber composites in 3-D printing and injection molded automotive products can improve fuel efficiency and emission standards among the US automotive vehicles.

633797 3D Printed Biocomposites from Biocarbon and Polyamide 12 Via Selective Laser Sintering

Abstract:

Additive manufacturing (AM) has revolutionized the plastic industry with its rapid prototype fabrication, part authentication and design flexibility. Selective laser sintering (SLS) is one such AM technique that is capable of manufacturing multiple complex shaped parts with no tooling or molds found with conventional extrusion-injection processes. As well, compared to fused deposition modeling (FDM), SLS has advantages in terms of its dimensional accuracy, complete infill, hollow or imbedded structures, throughput, self-supporting powder, and finish of the 3D printed parts. One of the downsides to SLS is the reusability of the un-sintered powder over many printing cycles, in which it degrades and can no longer be used. To offset the amount of polymer waste from SLS printing along with increasing the biobased content within the parts themselves, the use of sustainable biomass-derived carbon (biocarbon) had been added in increasing increments. The research focuses on formulations of polyamide 12 (PA12) powder mixed with biocarbon powder at ratios of 2, 5, and 10 wt.% inclusion. The biocomposites produced from SLS 3D printing were prepared into ASTM test specimens for performance analysis. The morphological, thermal, and mechanical properties were analyzed to determine the effect biocarbon had on the performance. The results demonstrated that with 5 wt.% biocarbon the impact strength was 33 J/m, with an increase in tensile and flexural strength by 4 and 21% respectively, while also improving the Young's modulus by 15% and flexural modulus by 19% over the neat PA12 SLS 3D printed sample.

Acknowledgements: The authors are thankful to: (I) Ontario Ministry of Economic Development, Job Creation and Trade ORF-RE09-078 (Project Nos 053970 and 054345); (ii) the Ontario Ministry of Agriculture, Food and Rural Affairs (OMAFRA), University of Guelph, Bioeconomy Industrial Uses Research Program Theme Project #030252 and 030485; the (iii) the Natural Sciences and Engineering Research Council of Canada (NSERC), Canada Research Chair (CRC) program Project No. 460788; and the NSERC Research Tools and Instruments (RTI) grant for their financial support.

638944 Evaluating durability after long-term thermal oxidation conditions of biocarbon filled biocomposites from polyphthalamide (PPA) and biobased polyamide (PA410) blend

Abstract:

All industries affected by climate change require significant changes to reduce their green house gases emissions (GHG), while at the same time they are currently experiencing societal pressures to decrease the environmental impact and increase the sustainability of their processes and products. Within the transport industry, a current strategy is to shift away from metal parts and replace them with plastic ones. In any application where the materials experience significant changes in temperature, their long-term conditioning must be evaluated. Polyphthalamide (PPA), a high-performance engineering plastic, is used in many different areas and is mostly derived from oil. In order to increase its sustainability, a composite with biobased polyamide 4,10 and biocarbon was processed (20 wt.% of biocarbon with 80 wt.% of a 25:75 PPA:PA410 polymer matrix). We measured the mechanical, thermal, thermomechanical and morphologic characteristics of the material, alongside its constituent parts and a similarly filled talc-composite, after thermal aging at 155 °C. All the materials were evaluated after every ~250 hours. The biocomposite became stiffer and brittle, with an increase in tensile and flexural modulus, and a decrease in impact energy. Infrared, dynamic mechanical analysis and morphological characterizations suggest that a decrease in chain length, as well as crosslinking between the remaining chains was responsible for the changes in the composite properties. Future work would require evaluating antioxidants that limit the thermal degradation without also affecting the mechanical properties.

Acknowledgements:

This research was financially supported by: (i) The Natural Sciences and Engineering Research Council of Canada (NSERC),-Collaborative Research and Development Grants (CRD) Project #401637 with the partner industries Prism Farms Limited and Competitive Green Technologies, Lamington, Ontario, Canada; (ii) the On-tario Ministry of Agriculture, Food and Rural Affairs (OMAFRA)/University of Guelph–Bioeconomy for Industrial Uses Research Program (Project #030332); (iii) the NSERC, Canada Research Chair (CRC) program Project No. 460788 and the Ontario Research Fund, Research Excellence Program; Round-7 (ORF-RE07) from the Ontario Ministry of Research, Innovation, and Science (MRIS) (Project # 052665 and 052662).

633848 Biodegradable Plastic Blends and Organically Modified Nanoclay-Based Sustainable Nanocomposite Films for Packaging Applications

Abstract:

Poly(3-hydroxybutyrate-co-hydroxyvalerate) (PHBV), a well-known biodegradable brittle polymer, was investigated with poly(butylene adipate-co-terephthalate) (PBAT) as a biobased toughener and various loadings of organically modified nanoclay as a reinforcement phase. This was conducted to determine the variation in mechanical, thermal and barrier properties after fabrication by two processes: (i) cast melt extrusion and (ii) compression moulding. The nanocomposite films were produced by combining PHBV/PBAT blend with 0.6, 1.2 and 1.8 wt% of nanoclay as shown in Figure 1(a). Improved oxygen and water vapour barrier properties were observed for the cast PHBV/PBAT nanocomposite films as compared to its compression moulded counterparts due to the uni-axial stretching of polymer chains. As a result, the melt casted nanocomposite films exhibited enhanced oxygen (~79%) (as shown in Figure 1b) and water vapor (~70%) permeabilities after adding nanoclay (1.2 wt%). This occurred because of the uniform dispersion of filler in the blend matrix as well as the strong adhesion between nanoclay and the blend matrix, which was confirmed by a rheological analysis. Another associated reason was the observed intercalation/exfoliation phenomena in nanoclay, which favorably occurred at the PHBV/PBAT interface, verified by a transmission electron microscopy (TEM). A further increment of clay content in polymer blends resulted in agglomeration problems. The cast PHBV/PBAT blend with 1.2 wt% nanoclay exhibited substantially enhanced (567.6%±0.1) elongation at break as compared to the blend matrix. This was caused by an improved stress transfer action, as compared with compressed films. Further, the formation of heterogeneous nucleation sites in both types of film were observed after adding nanoclay (1.8%) into the blend matrix.

Acknowledgements

The authors would like to thank the following for their financial support: Ontario Ministry of Agriculture, Food and Rural Affairs (OMAFRA)/University of Guelph-Bioeconomy for Industrial Uses Research Program (Project # 030054 and # 030255); the Natural Sciences and Engineering Research Council (NSERC), Canada Discovery Grants (Project # 400320); and the Ontario Ministry of Agriculture, Food and Rural Affairs (OMAFRA) – University of Guelph, the Bioeconomy Industrial Uses Research Program Theme (Project # 030255); and the Ontario Ministry of Economic Development, Job Creation and Trade ORF-RE09-078 (Project #053970).

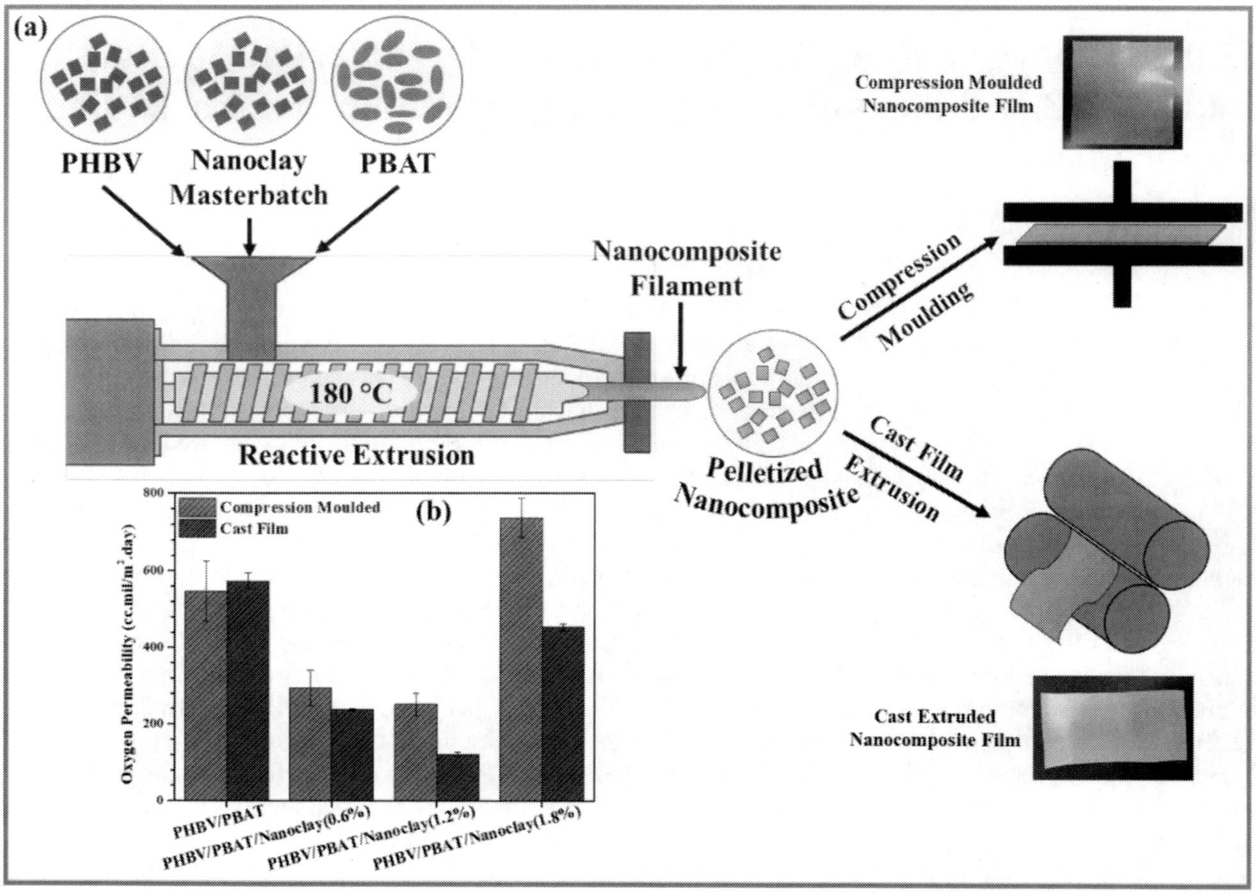
(a)
PHBV
Nanoclay Masterbatch
PBAT
180 °C
Reactive Extrusion
Nanocomposite Filament
Pelletized Nanocomposite
Compression Moulding
Cast Film Extrusion
Compression Moulded Nanocomposite Film
Cast Extruded Nanocomposite Film
(b)
Compression Moulded
Cast Film
Oxygen Permeability (cc.mil/m^2.day)
800
600
400
200
0
PHBV/PBAT
PHBV/PBAT/Nanoclay(0.6%)
PHBV/PBAT/Nanoclay(1.2%)
PHBV/PBAT/Nanoclay(1.8%)

633849 Biocarbon Production through Slow Pyrolysis of Chicken Feathers and Spent Coffee Ground Bio-Oil and Physicochemical Characterization of the Resulting Biocarbon

Abstract:

Pyrolysis of organic matter is a promising technique for recovering the value-added chemicals and energy from waste, as such it has encouraged extensive research within the last few decades. Energy obtained from biomass is considered carbon neutral, sustainable, clean and low-cost. The present study demonstrates the process of turning bio-oil (chicken feathers and spent coffee ground) into value-added biocarbon through slow pyrolysis techniques under an inert ambience. The biocarbon was obtained at three varying temperatures (400, 600 and 900 °C), at a 10 °C min^{-1} heating rate, for a 30 min holding time (HT) and was tested for its surface morphology, thermal stability, elemental composition, functionality, graphitic content, particle size, thermal and electrical conductivity. Further, the evolved gases during pyrolysis of bio-oil were also tested by a TGA-FTIR analyzer. The result of the TGA-FTIR study confirms that during bio-oil pyrolysis, hydrocarbons, CO_2, and carbonyl compounds are the major products. Pyrolysis results of bio-oil confirm that the operating temperature has a direct impact on the biocarbon yield and its properties. The physicochemical examination of gained biocarbon revealed that higher temperature-based biocarbon (600 °C and 900 °C) substantially enhances the thermal stability, thermal conductivity, graphitic content, and content of carbon. Elemental and Energy Dispersive Spectroscopy (EDS) examination of biocarbon established that higher temperature-based biocarbon (600 °C and 900 °C) has a lower amount of oxygen and hydrogen compared to lower temperature-based pyrolyzed biocarbon (400 °C). Also, the purest form of biocarbon was found at a higher temperature-based biocarbon with higher thermal stability and carbon content. The surface morphology of biocarbon confirms that higher temperature-based biocarbon (600 and 900 °C) provides larger and harder particles than lower temperature biocarbon (400 °C). Further, the electrical conductivity of biocarbon decreased; however, the thermal conductivity of biocarbon increased with a rise in the pyrolysis temperature. Moreover, the particle size analysis of biocarbon established that majority of biocarbon particles can be found in the range of 1 μm. The results from this study confirm that biocarbon can be used for various material applications.

Acknowledgements:

The present work is carried out with the support of: i) The Agriculture and Agri-Food Canada (AAFC) and Competitive Green Technologies, Canada through AgSci Cluster Program (Project No. 054712); (ii) the Ontario Ministry of Agriculture, Food and Rural Affairs (OMAFRA)/University of Guelph – Bioeconomy for Industrial Uses Research Program (Project Nos. 030331 and 030332); and (iii) the Natural Sciences and Engineering Research Council (NSERC), Canada Discovery Grants (Project No. 401111).

AUTHOR INDEX

AIChE
120 Wall Street, FL 23
New York, NY 10005-4020

ISBN 978-1-7138-5711-2

CURRAN ASSOCIATES INC.
proceedings.com